PEOPLE'S POWER

PEOPLE'S POWER

ASHLEY DAWSON

OR Books

New York · London

All rights information: rights@orbooks.com
Visit our website at www.orbooks.com
First printing 2022

Published by OR Books, New York and London

Library of Congress Cataloging-in-Publication Data: A catalog record for this book is available from the Library of Congress.

Typeset by Lapiz Digital.

paperback ISBN 978-1-68219-297-9 • ebook ISBN 978-1-68219-292-4

TABLE OF CONTENTS

CHAPTER ONE: INTRODUCTION

The Empire State Building was lit up emerald green on a chilly night in the winter of 2018 to celebrate New York City's announcement that it would be divesting its $5 billion pension fund from investments in fossil fuels. In tandem with this dramatic move to divest, the city also filed suit against the five biggest oil corporations, seeking reparations for the billions of dollars in damage that climate change has already inflicted on New York. It was a historic moment: The biggest city in the world's most powerful nation had come out against the planet's richest, most powerful, and most destructive industry. The divestment announcement was a landmark victory for the vast majority of the city's people against the fossil fuel industry, its well-heeled and powerful backers on Wall Street, and the globe-girdling military-industrial complex that supports contemporary fossil capital.

But it will take time for New York City's actions to have an impact. Divestment of the city's pension funds is not expected to happen until at least 2022. Also, Big Oil employs legions of lawyers who are adept at using the courts to obstruct justice and forestall reparations. Meanwhile, the Trump administration has unleashed policies of such environmental destructiveness that their impacts are likely to be measured in the geologic record, in degrees of temperature increase, and in feet of sea level rise around the world. By pulling the United States out of the Paris Agreement on Climate Change, and by aggressively expanding extreme forms of extraction such as hydrofracking, Trump has proudly proclaimed himself to be the embodiment of the Capitalocene, the age in which capitalism's relentless drive to expand has generated massive carbon emissions, pushing planetary ecosystems into states of unpredictable, deadly turbulence.[1] Of course, carbon emissions are collective and historical, so it would be wrong to suggest that Trump is solely responsible for planetary ecocide, but his election came at a critical juncture for the struggle to avert cataclysmic climate change. In declaring his intentions to unleash unfettered fossil capitalism, Trump epitomizes and promises to grievously aggravate the catastrophic contradictions of the Capitalocene.

While those who benefit from fossil capitalism might choose to ignore climate change, scientists are recording increasingly alarming trends in planetary systems. On the day Trump was elected, November 6, 2016, the World Meteorological Organization (WMO) released a report announcing that the past five years were the hottest ever recorded.[2] The WMO report also alluded to other grievous anthropogenic climate impacts, including rapidly rising sea levels, likely to surge in coming years as a result of the unexpectedly rapid melting of polar ice. As delegates gathered in Marrakech the week after the 2016 election for the twenty-second annual UN climate summit, leading scientists warned that global temperatures are now two-tenths of a degree away from the upper threshold agreed upon only one year previously during the Paris negotiations.[3] Researchers also warned that the Arctic is experiencing extraordinarily hot sea surface and air temperatures, which are stopping ice from forming at the North Pole and intensifying a warming feedback effect.[4] As if these doomsday trends were not enough, the World Wide Fund for Nature reported at the beginning of the month in its *Living Planet Report* that global populations of fish, birds, mammals, amphibians, and reptiles declined by 58 percent between 1970 and 2012, and that the planet is on track to lose two-thirds of its sentient life forms in the fifty-year period ending in 2020.[5] In 2019, the Intergovernmental Panel on Climate Change released a series of "doomsday" reports that added to this apocalyptic drumbeat. Terms like "global warming" and "climate change" are far too anesthetic to characterize the crisis of our times: We are living through a planetary environmental cataclysm that is on track to exterminate most life on Earth.

The climate justice movement that won such a glowing victory with the announcement of New York's divestment plans is driven by one clear imperative: the cities and nations of the world must cease burning fossil fuels if we are to avert looming planetary ecocide. Though it has many facets, the climate crisis is above all an energy emergency. The energy sector is responsible for at least two-thirds of all greenhouse gas emissions.[6] Hundreds of thousands of pipelines, wells, drilling platforms, and mines snake through and across the planet, driving an extractive economy whose destruction is justified by increasingly pervasive authoritarianism, often facilitated by bloody violence.[7] The world is growing increasingly hungry for power, thus carbon emissions continue to rise despite every effort to rein them in. Given our current reliance on fossil fuels, more demand for power necessarily means more dangerous climate chaos. According to the International Energy Agency (IEA), overall global energy demand is projected to accelerate significantly in the coming decades, expanding by roughly one-third between now and 2040.[8] This inconvenient truth has generated a sense of urgency around the need for an energy revolution, a wholesale shift from dirty fossil fuels to clean renewable energy.

The great task of our times is to stop all new fossil fuel infrastructures. All of our other efforts to fight climate change will be useless if the world does not transition away from fossil fuels in the next decade or so. If we wish to prevent catastrophic climate disruption, we cannot dig up more coal, drill for more oil and gas, and further exploit tar sands. And we must also stop building pipelines and export terminals to transport freshly extracted fossil fuels. In other words, although some portion of currently existing reserves must be used to power the transition to renewables, we have to stop prospecting for, extracting, and burning fresh supplies. This is because the carbon stored in *currently* operating fields and mines is enough to push the atmosphere above a critical threshold, generating warming of more than 2° Celsius and unleashing catastrophic feedback effects.[9] We've already melted half of the Arctic ice and let loose some of the most ferocious droughts and floods in human history, and that's after only 1° of warming. To meet the 1.5° goal agreed to in Paris and keep the planet habitable, we must shut down *all* the world's coal mines and some of the oil and gas fields currently in operation before they are exhausted.[10] And this must happen quickly: the longer we tarry, the more wrenching the transition will necessarily be, and the greater risk we run of triggering unstoppable feedback effects in planetary ecosystems that catapult us toward planetary ecocide.

The movement for climate justice has developed a range of innovative nonviolent direct-action tactics, as well as an expanding array of legal and political strategies, to block pipelines and other fossil fuel infrastructures. The movement has won some important victories, but as I will show later in this book, we are battling a global bonanza of extreme fossil fuel extraction. If humanity is to prevail in this fight for a habitable future, we must be clear about what the obstacles are that prevent a transition beyond fossil fuels. The climate justice movement must set forth clear and emancipatory alternatives to fossil capitalism. And we must be clear that our struggle is not just about remaking our energy infrastructures, but also about transforming the cultural outlooks and institutions that fossil capitalism has generated.[11] Fossil capitalism has shaped societies across the globe in ways that are both material and cultural, generating tangible things like highways and plastics, as well as intangible habits, beliefs, and even affects.[12] If we wish to transform the material infrastructures of fossil capitalism, we must also interrogate and even upend the structures of feeling they have helped produce.[13]

The trouble is that cheap and abundant fossil fuels are the basis for widely accepted expectations of unbridled freedom, ceaseless growth, and spiraling consumption in the world's developed nations. To be seen as a winner—and sometimes simply to survive—in today's hyper-competitive world is to be able to participate gleefully in time- and space-annihilating patterns of mobility and consumption fueled by the combustion of ancient sunlight.[14] Such fossil-fueled lives are the acme of aspiration for many people around the world. We live, in other words, in a global carbon culture, and are the inheritors of a worldview thoroughly saturated by fossil fuels, even if the natural resources that power this cultural outlook tend to remain inscrutable if not entirely invisible to most of us.[15] Prescriptions for the transition beyond fossil fuels offered by mainstream commentators seldom highlight, let alone challenge, these structures of fossil feeling. All too often, for example, questions of energy policy are treated as purely economic or technological issues, making them almost as mysterious to the average person as the origin of the electricity that powers their phones and computers. In the face of the Trumpocene, for example, pro-market environmentalists have pointed to underlying economic trends, such as the increasing cheapness of solar power, to suggest that the world has embarked on a transition to clean energy that is unstoppable, notwithstanding Trump's extravagant promises to the fossil fuel industry. For prominent advocates of green growth like former vice president Al Gore, the world is on an irrevocable pathway toward decoupling economic growth from carbon emissions. This green capitalist view has percolated through entities like the World Bank, which has predictably embraced "inclusive green growth." But it has also impacted environmental organizations such as Greenpeace, which concluded after the twenty-first

UN Climate Summit (COP21) in 2015 that "the end of fossil fuels is near, we must speed its coming."[16] For green capitalists, the transition from fossil fuels to renewable energy is well under-way and will not disrupt any of the fundamental expectations of capitalist modernity.

Unfortunately, this optimistic outlook is based on wishful thinking. The United States gets less than 10 percent of its energy from renewable energy sources like wind and solar at present,[17] but leading climate scientists argue that renewables need to make up at least 30 percent of the global electricity supply by the end of 2020 in order to meet the goals of the Paris Climate Agreement.[18] As climate experts note, "Timing is everything . . . should emissions continue to rise beyond 2020, or even remain level, the temperature goals set in Paris become almost unattain-able."[19] In addition, the optimistic predictions of green capitalists reflect partial information about the global energy economy that obscures the fact that we are in an age of rapidly expanding fos-sil fuel extraction.[20] Among the trends fueling the optimism of pro-market environmentalists are a sharp drop in coal consump-tion, rising investment in renewable energy, slowing energy demand and improving energy intensity, and the leveling off of global carbon emissions. Taken together, these trends are read to signify the decisive deterioration in the economics of fossil fuels *as a result of* the strengthening economic position of renewables. But this is a fundamental misreading.[21] As data from organiza-tions such as the International Energy Agency (IEA) shows, while coal production may be down in the United States, global coal use has *doubled* since the mid-1980s. Similarly, natural gas use is increasing not just in the United States, but also on a global scale, with energy generated from gas growing at a faster rate than that generated by renewables. Oil consumption is also increasing on a global scale. According to recent assessments, so-called modern renewables still only generate 10.3 percent of global electrical power.[22] But this statistic rather deceptively includes 3.9 percent hydropower, which is not categorized as a modern renewable since big dams generate significant carbon emissions from the forests they drown. Modern renewables like wind and solar energy combined actually only generate 6.4 percent of global electrical power. The percentage of renewable power in the transportation and heating and cooling sectors, which together account for 80 percent of global final energy demand, is far more miniscule. In these sectors, energy transition has barely begun. Fossil fuels still account for 78.3 percent of global energy consumption.[23] And despite a diminishing rate of global carbon emissions in recent years, aggregate emissions continue to increase, pushing atmospheric carbon concentrations and global warming across dangerous thresholds toward runaway climate chaos. In other words, while we might be witnessing a glut of fossil fuel production that has driven prices down to historic lows and made some industries, such as coal, uncompetitive in some nations, the fossil fuel era is emphatically not over. Although renewable energy production has certainly been increasing, it has not been growing fast enough to displace fossil fuels, and it will not do so without decisive political action to shift the world toward a just transition. Current market-based efforts to shift to renewable energy are simply not working at the necessary pace and scale.

We need a rapid and just transition beyond fossil fuels. We are only going to get such a transition if we can wrench control of our energy systems out of the hands of profit-seeking corporations with a strong stake in continuing business as usual. The struggle for democratic control over energy production, distribution, and use is consequently the key front in the fight for a better, sustainable world. It is not simply that it's taking too long to make the transition. The problem is a more fundamental one. The capitalist system that orients current efforts to shift energy regimes is based on a fundamental logic: grow or die. This growth imperative is a recipe for the mass extinction of most species on a finite planet.[24] As long as energy production remains grounded in the logic of the capitalist market, it will continue to obey capitalism's fundamentally irrational drive to generate ceaselessly expanding profits through endless growth. If capitalists can make money off fossil fuels, they will continue to drill holes in the earth, the environmental and social consequences be damned. Since fossil fuel corporations have immense amounts of assets sunk into existing infrastructures, they will fight to prevent a transition to a zero-carbon society—despite the occasional charade of moving "beyond petroleum." Truly sustainable energy production will only be possible if power is taken out of the hands of these gargantuan profit-seeking corporations and their flunkys in the halls of state. Ordinary citizens and communities must consequently control power, in both senses of the term. Decisions about energy generation and the transition to renewable energy need to be oriented around genuine collective needs and framed by a horizon more ample and more sane than the nihilistic, short-term perspectives of the current capitalist system. This transition also must include shifting workers in current fossil fuel industries to well-paying green jobs, generating a kind of security in employment that the free market will never guarantee.[25] If our collective future is presently being determined by a small cabal of fossil oligarchs, who are making money while driving the planet toward biological annihilation, the alternative to such folly is to collectivize the control of energy and to organize a just transition to renewable power through participatory, democratic control.

In order to make this power shift, we need to stop thinking of energy as a commodity and instead conceive of it as part of the global commons, a vital element in the great stock of air, water, plants, and collectively created cultural forms like music and language that should be regarded as the inheritance of humanity as a whole.[26] The commons are made up of material things—those tangible, finite resources such as clean air and water upon which all life depends. But the commons also consist of intangible, nonfinite collective resources such as knowledge, shared customs, means of communication, and even more ineffable things such as collective emotions like joy and anger, all of which might be termed the social commons. Energy needs to be thought of as both sorts of commons since it is composed of both the "natural resources" (coal, oil, gas, wind, sun, tides, etc.) from which power is generated and of the technologically and socially distributed power derived from these resources. That energy ought to be thought of as a common good—literally, as common wealth—is clear when one scrutinizes its sources, the product of their social use, and the urgent need for a just transition to renewable energy.[27]

In terms of the sources of energy, it should be quite clear that both fossil fuels like coal and oil and sources of renewable energy like the wind and the sun are part of a natural commons. They are physical resources (in the case of fossil fuels, finite ones) that should be shared equally throughout society. The fact that they presently are not equally distributed, and instead are hoarded and exploited for profit by the few, reflects the fundamental injustice of the capitalist system and the forms of warmongering and imperialism that it has historically produced. But the idea of energy as a common good is not as outlandish as it might at first seem. After all, if citizens of wealthy nations have grown accustomed to thinking of fossil fuels in a commodified form, as the bill paid at a local Exxon station or the monthly charge from a regional power utility, it was not always so. Within the United States itself, but even more so among people fighting domination and injustice in other parts of the world, the idea of fossil fuels as the common property of the people—a People's Power—has resonated powerfully and helped spark radical social movements. Given capitalism's tendency to generate unequal access to and scarcity of the natural commons, inequalities with deadly implications, it is only to be expected that strong countervailing movements asserting collective control over this common wealth should arise in protest. *Power to the People* is a resonant rallying cry to the world over, particularly since it seems to condense ideas relating to equal access to the energy commons with the desire to level today's outrageous social inequalities. In addition, if the commodification of fossil fuels, whether by global megacorporations or their powerful state-based analogues, has become a fait accompli today, this skewed and inegalitarian situation has not yet become commonsense in relation to renewable energy sources. It seems patently absurd, in fact, to think that a corporation or a state would lay exclusive claim to the sunlight or the wind.

To fully realize the implications of treating energy as a commons, though, we must challenge the deeply ingrained idea that energy is a thing: a joule or kilowatt-hour. This goes against the grain of contemporary conceptions of energy as an abstract biophysical property governed by the immutable laws of thermodynamics. But this objectifying meaning of energy only developed in the eighteenth century, at the inception of the era of fossil capital. Prior to this, energy was a far more flexible term, one that tended to refer to a kind of vital force produced under specific circumstances. The word "energy" in fact derives from the Greek words *en* (within) and *ergon* (work).[28] Aristotle developed the term, using it to denote the active capacity of the human intellect.[29] By the late seventeenth century, when the German Enlightenment thinker Gottfried Leibnitz rendered the term in Latin as *vis viva*, or "living force," energy had come to refer to the internal vigor required by a person to engage in physical or mental activity. This usage was appropriate for an age in which the transatlantic slave trade was turning millions of commoditized human beings into a hitherto unrivaled source of energy, a power so potent that it was used to transform entire regional landscapes in the Americas and to provide the energy to kick-start the Industrial Revolution.[30] The energy commons thus includes brutal histories of human subjugation and exploitation, as well as ideas of shared access to "natural resources." Today's fight for the energy commons must reckon with the lingering effects of these histories of settler colonialism and racial slavery. The energy commons therefore must be about more than simply switching from fossil fuels to solar power: at its heart, this struggle must enable radical redistributions of power that don't just democratize but also effectively decolonize energy and society.

As crucial as the fight against new pipelines and other fossil fuel infrastructure is, in other words, the struggle against fossil capitalism must also have a positive content: the means of energy production must be socialized and collectively controlled, if we are to avoid perpetuating the glaring social inequalities of fossil capitalism within a renewable energy regime.[31] The idea that the planet can be saved if we simply swap solar energy farms for coal-fired power plants maintains the toxic illusion that it is possible to sustain current levels of energy consumption, levels that expand relentlessly.[32] Indeed, as renewable energy is increasingly adopted, it is likely to be subjected to the same profit and growth imperatives that characterize energy under fossil capitalism. But even if energy is generated from 100 percent renewable sources, the transition to a new, renewable energy regime will not by itself solve the climate crisis since it is only responsible for a portion of emissions. This idea might seem farfetched given the utopian rhetoric that surrounds forms of renewable energy such as solar and wind power, but it is worth remembering that capitalism existed for at least two centuries before fossil fuels were integrated into its mode of production. The racial capitalism of plantation slavery used solar and wind power as well as human energy to grow sugarcane and to transport slaves and sugar across the Atlantic.[33] Capitalist exploitation, imperialism, and genocide have existed under renewable energy regimes in the past, and may exist in the future unless energy is conceived of as a social commons to be collectively governed for the good of all, unless the transition to renewable energy in terms of a broader power shift toward egalitarian, noncapitalist social relations is conceptualized.

Against this dystopian future, the global movement for energy democracy is fighting not simply to shift to renewable energy but to bring power under public and community ownership or control. But the new forms of public power should not be seen as a revival of the top-down, centralized, and bureaucratic models of the past. Critics have skewered the public services of social democratic states for being undemocratic and inefficient.[34] This criticism was correct at times, if not always. Yet the notion that private ownership constitutes a viable antithesis, that the free market holds the key to innovation, enterprise, and even democracy, has proven disastrously misguided and damaging. Often known as neoliberalism, the ideology that fetishizes unbridled capitalism was part of a conservative counterrevolution that was often imposed by force rather than through market efficiency.[35] We live among the tattered remnants of public infrastructure created during the mid-twentieth century: we are bedraggled denizens of a world of vertiginous, highly concentrated private affluence set amidst public destitution and squalor. The critique of public services has led to precariousness and rampant injustice. It is time to reclaim the common good and public power. The energy transition must be about new forms of collective commoning.

This reclamation is given additional urgency by the fact that we live in a time of deepening popular discontent with elite discourses about globalization and the wonders of the free market. Relentless rounds of austerity have thoroughly hollowed out such ideas. But today, a new round of populist demagogues who are taking the conservative counterrevolution in an increasingly authoritarian direction has harnessed popular anger over the dysfunctions of neoliberalism. From Trumpism in the United States to Bolsonaro in Brazil and beyond, xenophobic nationalist forces are exploiting economic stagnation, rising inequality, and the increasingly apparent impacts of climate chaos. Movements like Occupy and the Arab Spring that called for economic and political democratization earlier in the decade have given way to strongmen like Trump, Bolsonaro, and the Philippines' Rodrigo Duterte, who scapegoat immigrants and the poor for their country's economic struggles while further privatizing the public sphere, cracking down on dissent, and expanding planet-destroying extractive industries. If popular discontent with neoliberal globalization is not to be even more thoroughly captured by the forces of reaction, we urgently need to articulate viable models of public ownership based on principles of economic egalitarianism, decentered decision-making, and public participation that reanimate the idea of democracy and the common good.[36]

We need new narratives of public potential that inspire a sense of radical possibility and progressive popular struggle.

In no realm is this reassertion of public prerogatives more important than in the struggle for energy democracy. Our collective survival literally hinges on this fight. The current centralized model of energy, a legacy of the age of fossil fuels, is characterized in the United States by the control of energy generation and distribution by a small number of large, for-profit corporations. Concentrating power and wealth in the hands of a few, this centralized form of power is the model of choice for corporate renewable energy transition.[37] But renewable energy resources are inherently distributed. Rather than generating electricity from one central power plant, that is, renewable energy comes from solar, wind, and geothermal resources—as well as energy efficiency, storage, and conservation—that can be found and harvested everywhere. The movement for energy democracy is thus fighting for decentralized forms of power in both senses of the term: energy that is generated locally, and energy that takes power away from today's economic, political, and energy oligarchies in order to empower the frontline communities—working people, indigenous groups, communities of color—who bear the brunt of fossil capitalism's toxic environmental, economic, and repressive elements.

Fighting to devolve power and decision-making should not, however, be seen as a dogmatic commitment to localism. New forms of direct democracy and decentralized power will only be able to flower if movements for the common good win control of the state at various levels and scales. Community control of renewable energy needs a progressive regulatory framework on an urban, a regional, and ultimately a national scale if we are to make the transition to renewable energy with the speed and scale necessary to avert planetary ecocide. Equally important, considerations of equity dictate that we cannot leave the rich to build resilient renewably powered microgrids while the rest of us are stuck with increasingly expensive and threadbare electricity provided by for-profit utilities. In addition, roughly one billion people around the world still lack access to electricity; our collective future hinges on finding ways to eliminate this global energy poverty without expanding fossil fuels.[38] How can we transform the existing public utilities—whose present form is a reflection of popular struggles against monopoly power as well as the containment of such struggles in the past—while also supporting decentralized power and self-managed energy provision? This struggle in the realm of energy reconnects us with a long history of battles to extend the political freedoms in existing institutions of representative democracy while also growing and multiplying new forms of direct democracy.[39]

The book you hold in your hands provides you with an overview of contemporary struggles around the global energy commons. Although it is intended ultimately to inform activism aimed at securing democratic control over this common wealth, the book begins from the conviction that we need to understand the ways in which energy infrastructure is embedded in much broader cultural and political force fields. The second chapter of this book therefore begins by discussing the ways in which neoliberal dogmas about the free market shape contemporary energy systems. Investor-owned utilities (IOUs) have been the primary providers of electricity to most citizens of the United States for the last century, indeed, since the establishment of the modern electrical grid. Yet today, they are caught in a death spiral as more renewable energy comes online. In this chapter, I explain why modern renewables like solar and wind power are such a threat to IOUs, and how the IOUs have been obstructing the transition to renewable power. The chapter places this resistance to change in the context of the fracking revolution, which has generated a massive upsurge in the production and consumption of fossil fuels. I explain here how policymakers adopted subsidies for the financial industry during and after the Great Recession of 2008 that ended up driving a fracking bonanza. This shift, I argue, constitutes the financialization of fossil capitalism. The chapter explains why this shift has made the increasing economic competitiveness of renewable energy inconsequential: it constitutes nothing more than an addition to an energy system already engaged in frenetic overproduction.

Chapter 3 of *People's Power* explores the history of struggles for electrification in the United States in the first third of the twentieth century. Today, as we embark on a momentous struggle for transition to clean power, we face many similar social and political questions to those confronted by key protagonists in the battle to electrify the United States in the 1920s and 1930s. For protagonists of this fight, such as the conservationist Gifford Pinchot, who led the campaign to transform the state of Pennsylvania's power infrastructure in the 1920s, the coming of electricity promised to contribute to new relations of production and exchange based on social solidarity rather than exploitation. Similarly, a decade later, key figures in the New Deal campaign for rural electrification fought for an energy transition that would not only link the nation's millions of farmers to the grid, but, in vanquishing the power of the monopolistic power utilities, would help spark a broader transition toward more egalitarian and genuinely democratic social relations. If past energy transitions ended up consolidating and even strengthening the power of elites, what do the efforts to catalyze a different kind of transition in the midst of the Great Depression have to tell us today as we seek to take advantage of the unique opening provided by the transition to clean power to transform our plutocratic and violently unequal society—and save the planet to boot?

The subsequent chapter explores the extent to which ideas of the commons can serve as an inspiration for contemporary struggles for energy democracy and just transition. The chapter shows how received ideas about a "tragedy of the commons" ironically find their clearest expression in the history of fossil capitalism as a result of the legal structures established to govern extraction. The juridical order of fossil capitalism produced rampant over-exploitation that necessitated both informal and state-orchestrated means of management of common resources. After documenting this history, the chapter discusses efforts to nationalize oil resources in countries such as Mexico and Iran, arguing that energy worker–catalyzed uprisings were predicated on ideas about oil as common wealth. All too often, these uprisings, when not put down viciously by imperial forces, could not escape the corrupting impact of oil and oligarchy. Nonetheless, these efforts are important examples of the fight to democratize control of energy resources. To what extent can "modern renewables" like solar and wind power escape the "resource curse"? This chapter investigates the impact of renewable energy's relatively accessible character. Since the wind blows and the sun shines everywhere, all communities may access renewable energy. Making sure that powerful individuals and corporations do not assert control over such relatively accessible resources, I argue, will hinge on the establishment of legal arrangements ensuring that they are constituted as collective resources. The chapter therefore concludes with an examination of legal histories and theories of the commons. The commons needs to be seen not simply as a resource constantly threatened by commodification within the capitalist system, but also as a counterforce being actively created by communities around the world.

Chapter 5 travels to Europe to look at the fight for energy democracy and just transition. Germany provides a particularly inspiring example of successful efforts to gain popular control over the energy commons, whether through small energy collectives or through re-municipalization of energy provision in cities like Hamburg and Berlin. Germany's famous *Energiewende*, or "energy transition," offers inspiration to activists in other nations, but the challenges faced by movements for energy democracy—not to mention for access to power in the first place—in places like the United States and in the nations of the Global South are different and substantial. Will efforts to make a market-led transition to renewable energy in the context of large investor-owned utilities like those in the United States be successful? What alternatives are grassroots environmental justice organizations in cities like New York experimenting with, particularly now that Trump and the Republicans have made progressive legislation impossible on the federal level?

Finally, the conclusion considers the implications of a world oriented around sun-derived power. While fossil fuels are themselves ultimately based on solar energy, they come to us in extremely concentrated forms, constituting a unique bonanza for the portion of humanity that has harnessed this power over the last several centuries. We must move into a future without such concentrated resources of energy. While doing so, we also need to triple the amount of energy being generated since we need to electrify everything, including all transportation and the heating and cooling of the buildings we inhabit. The challenge here is immense, but luckily, earth is flooded every day by tremendous amounts of solar energy. The prospect of a harmonious future powered by such practically unlimited power has given rise to new genres of speculative fiction such as "solarpunk," an example of which I will discuss in the conclusion. But while solar radiation might be practically unlimited, the terrestrial resources we must use to harness it—from steel to lithium to rare earth elements—are not equally abundant. How are we to avoid transitioning to a world powered by renewable energy but characterized by new forms of capitalist exploitation, extractivism, and imperialism? The ways in which we imagine the future, I argue, are key to today's struggles for just transition.

This book is written from the foundational conviction that energy production must not pollute the environment or harm people, but that everyone must have access to adequate energy. *People's Power* argues that the free market will not meet either of these key goals. Only a radical movement for energy democracy that rejects the ceaseless expansionist imperative of capitalist culture will ensure that fossil fuels are left in the ground, that no one is left behind as we transition to renewable energy, and that global energy poverty is addressed. The only way to win the struggle for an egalitarian and sustainable civilization upon which our collective survival depends, *People's Power* suggests, is to socialize and democratize the means of energy production. Hand in hand with the assertion of genuine public control over energy production and distribution must be a decisive repudiation of the feckless attitudes toward energy consumption that have characterized the age of fossil capitalism. This repudiation does not mean an embrace of sweeping austerity for the many and cloying affluence for the few. Instead, the idea of an energy commons holds out the promise of new forms of social connection and solidarity as we collectively steward the common wealth of which energy is such an integral part. In fighting for democratic control of the energy commons, we struggle for a new era beyond fossil capitalism, one based in diversity, autonomy, and equality rather than today's current world of stark exploitation and spiraling degradation.

CHAPTER TWO: THE FOSSIL CAPITALIST DEATH SPIRAL

It is not often remembered that Barack Obama began his campaign for president with an invocation of FDR's mobilization of American industry in the war against fascism. Speaking at the Detroit Economic Club in the spring of 2007, Obama recalled to the auto industry executives and other business leaders in attendance the moment when Detroit became the "Arsenal of Democracy."[40] When his advisors told President Roosevelt that his production goals were unrealistic, Obama recounted, he waved them off, saying that "the production people can do it if they really try." Obama told his listeners that Detroit had become known as the "Miraculous City" after the auto industry and its workers shifted from producing cars to planes, tanks, and arms, in the process becoming one of the nation's most important contributors to the war effort. This remarkable mobilization, Obama argued, is "the kind of American miracle we need today." Obama's speech was undeniably audacious in its awareness of the magnitude of the threat posed by fossil capitalism: "At the dawn of the twenty-first century," he declared, "the country that faced down the tyranny of fascism and communism is now called to challenge the tyranny of oil. For the very resource that has fueled our way of life over the last hundred years now threatens to destroy it if our generation does not act now and act boldly."

In a pattern that would become familiar in his years in the White House, Obama went on to propose a series of measures to cope with carbon emissions that fell far short of his stirring opening invocation of wartime mobilization to stave off the collapse of fossil-fueled civilization. The two key initiatives he announced during that speech in Detroit were higher fuel efficiency standards for cars and a national low-carbon fuel standard—both laudable policies but neither commensurate with the scale and severity of the threat he himself depicted at the outset of his speech. Later in his campaign, Obama talked up an ambitious "blueprint for a green economy," sparking hopes of a Green New Deal in the form of an economic stimulus plan—the American Recovery and Reinvestment Act—that would drive investment in renewable energy and green infrastructure.[41] The key element of Obama's inspirational message was a pledge to transition workers to a low-carbon economy, in the process creating five million new green jobs. But although the stimulus plan did direct a big infusion of cash into the green energy and technology sectors—providing seed money, for example, for solar generation projects and wind farms nationwide[42]—these projects were not enough to offset job losses resulting from the Great Recession.[43] When Congress began to throttle back the stimulus, private capital did not flood in as predicted and jobs failed to materialize. By 2011, only 3 percent of workers were involved in "green goods and services."[44] According to Jeremy Brecher of the Labor Network for

Sustainability, the failure of Obama's promise to establish a green economy left organized labor feeling bitter and betrayed.[45]

Labor was not alone in feeling dismayed by the gap between Obama's rhetoric and the policies his administration advocated to address the climate crisis. Environmental justice groups like Green For All backed the stimulus plan based on the idea that it would help bring good jobs to communities hard hit by the Great Recession. But these communities did not see significant economic revitalization since workforce programs aimed at low-income and disadvantaged communities were never scaled up to levels significant enough to offset racial disparities in joblessness.[46] These concerns were increased by the failure of Obama's Clean Power Plan to address questions of equity. In mid-2015, the Environmental Protection Agency, using the authority of the Clean Air Act, proposed a set of rules that would force states to cut their carbon emissions. The targets for cuts were relatively modest given the enormity of the climate crisis, and were not slated to come into effect until 2022—giving the Right plenty of time to roll back the plan, a step that the Trump administration took almost immediately.[47] Perhaps most galling, however, was the fact that the Clean Power Plan failed to mandate emissions reductions for environmental justice communities, failed to prioritize energy efficiency and renewable energy deployment in those communities, and allowed states to use carbon trading to fulfill their obligations under the plan, a policy that environmental justice groups firmly opposed.[48]

Notwithstanding the criticism he received for his efforts to address the climate crisis, Obama remained buoyant. In an article published in the journal *Science* shortly before he left office, Obama boldly declared, "I believe the trend toward clean energy is irreversible."[49] There are four reasons, Obama argued, to believe in the irreversible momentum of clean energy: the decoupling of economic growth and energy-sector carbon emissions in the US economy during the previous seven years; emissions reductions in the private sector during this period, as businesses embraced energy efficiency in order to benefit the bottom line; the impact of market forces in the energy sector, as natural gas replaced coal and renewable energy became increasingly cost competitive with fossil fuels; and, finally, the global competition to go green generated by the Paris Accords.[50] Although Obama published his upbeat assessment seven days before Donald Trump was sworn into office, Trump's determination to roll back Obama-era efforts to address the climate crisis was no secret. Obama's article in *Science* should therefore be seen as an assertion that Trump would not be able to unravel the fundamentals of the already-in-progress energy transition. But was Obama engaging in wishful thinking about the irreversible character of his legacy? Are we really in the midst of a massive transition in global energy systems? Are solar and wind power replacing dirty fossil fuels like coal? If so, why do greenhouse gases continue to build up in the atmosphere? Will cheap clean energy save us from climate change?

Granted, substantial strides have been made by renewable energy in recent years, but the pace of the market-led transition is too slow to avert planetary ecocide. Major technological, economic, and political obstacles must be overcome if the energy sector is to be decarbonized successfully in the vanishingly limited amount of time we have to avoid cataclysmic climate change. Thus, if the energy transition is to be accelerated to the degree that science tells us it must be, the idea that energy transition is something that can be left to the free market and to technical experts must be challenged. We need to revive the sense of urgency that Obama articulated during the early stages of his bid to become president and reanimate the vision of a wholesale political, economic, and cultural mobilization in order to address the climate crisis. Yet even if the great task of our times is to stop all new fossil fuel infrastructures, winning a transition to renewable energy will not be sufficient to avert climate chaos. Unless we dismantle and replace a capitalist system based on extreme extraction, inexorable growth, mounting inequalities, militarism, and colonialism, our headlong rush toward extinction will continue. We need not just decarbonization, but global system change.

Unfortunately, many advocates of renewable energy are fervent believers in a market-led shift toward clean power. For Michael Liebreich, chair of the advisory board of Bloomberg New Energy Finance, for example, Donald Trump's efforts to revive fossil capitalism through his *America First* energy plan don't account for the economic realities of coal and renewable energy.[51] Liebreich points out that Trump's summary dismissal of wind and solar power as expensive is simply inaccurate: renewables now outcompete coal economically. These facts lead Liebreich to conclude that "the world will continue its inevitable transition to clean energy and transportation [no matter what Trump does in office], just at a slower rate than if the US were fully committed to leading the process."[52] Liebreich is not alone. His celebration of an inexorable, technology- and market-led transition to renewable energy is echoed by prominent figures like former vice president Al Gore and former UN climate chief Christiana Figueres, who reacted to Trump's threat to withdraw from the Paris Agreement by saying, "There is a big difference between the economics of climate change and the politics of climate change. Is Trump going to stop that advance [by businesses toward low-carbon technologies]? I don't think so."[53]

The ideological significance of these conclusions about a market-based transition to renewable power is momentous. If the free market really is autonomously ending fossil fuels, no organized opposition to the Trump agenda is necessary.[54] According to this logic, the struggle for energy transition can be fought purely in the realm of individual consumer decisions. We vote with our wallets, as atomized private individuals rather than as citizens of a republic in which the common good is articulated through public debate and social contestation. In a consumer nation, we trust in the dual fetishes that have dominated the American twentieth century—the free market and technological innovation—to save us from environmental oblivion.[55] By continuing to believe in the shibboleth of the free market, the unstated credos of green capitalism not only shape our collective imaginary about how energy transition will occur, but also dictate who will benefit from this transition. A green capitalist transition would be one in which existing economic and political relations were not significantly unsettled, one in which public power does not rein in corporate power and moneymaking prerogatives, and one in which the fundamental orientation of the capitalist system toward ceaseless, unchecked growth is not challenged. We should be clear, in sum, that arguments about an energy transition based on the free market are deeply ideological, and have profound social and economic stakes. If they are correct, we may eventually make a successful energy transition, but existing social relations will not be significantly transformed. If they are incorrect, however, not only will existing social oppression go unchallenged, but we will also continue on our present terrifying path toward an environmental breakdown.

In his influential 2007 climate crisis report, the economist Nicholas Stern famously argued that "climate change is the greatest market failure the world has ever seen."[56] This admission, that the unfettered free market is responsible for the climate crisis, is quite startling coming from a former World Bank vice president. Historians of capitalism agree with Stern's assessment. As Karl Polanyi argues in his seminal book *The Great Transformation*, "[The] idea of a self-adjusting market implied a stark Utopia. Such an institution could not exist for any length of time without annihilating the human and natural substance of society."[57] Yet if a totally unregulated market would destroy humanity and the planet, for Polanyi, capitalist markets were, from their inception, a result of political machinations: The "laissez-faire economy," he argued, "was the product of deliberate State action."[58] The idea of a self-regulating free market is an illusion, a toxic one at that. While most societies across history have included some sort of markets, none—our global capitalist one included—has been ruled absolutely by market logic. As Polanyi put it, "The road to the free market was opened and kept open by an enormous increase in continuous, centrally organized and controlled intervention."[59] The dominance of neoliberal ideology in recent decades has helped obscure the pivotal role of the state in stimulating and propping up particular sectors of capital around the world. In fact, many of the recent technological innovations that are supposed to demonstrate the entrepreneurial dynamism of the private sector, from the internet to voice recognition and touch screen technology, were in fact developed with state-funded research.[60]

It is particularly ironic that liberal advocates of renewable energy should place their faith in an energy transition driven by the free market. Renewable energy confronts anything but a level playing field. We are in the midst of a global energy war, one in which markets are shaped by state power largely captured by and subordinated to the interests of fossil capital. Governments around the world provide $775 billion to $1 trillion annually in subsidies to support the oil, gas, and coal industries, a figure that doesn't include costs for environmental and health impacts, military conflicts, and climate change.[61] When such "externalities" are included, as in a 2015 study by the International Monetary Fund, the unpaid cost of fossil fuels tops $5 trillion annually.[62] Energy resources today are still largely controlled by a fossil capitalist system whose devotion to ceaseless growth threatens to annihilate most life on this planet. To hope for a market-led technological fix is folly. An energy transition will not take place with the speed necessary to avert planetary ecocide, unless the links between the fossil fuel industry, its unscrupulous financiers on Wall Street, and its enablers and enforcers in government are exposed and broken. Energy resources must be wrenched out of the hands of fossil capitalism, which inevitably will involve confronting and overthrowing the control of the state by the fossil fuel industry. The struggle for energy transition is thus a fight for public and collective control of energy resources, and for democratic control of the state power that shapes the development of such resources. It is, in sum, a struggle for energy democracy.

TWILIGHT OF THE IOU

Sunny Hawaii has a history of embracing renewable energy. Today, over 10 percent of households in the Aloha State sport rooftop solar systems to generate electricity, a figure more than double the proportion in the famously pro-renewable state of California.[63] But back in 2013, Hawaii became a victim of its own success in the installation of renewables: the rapid increase of privately installed solar panels overloaded the state's power grid, leading the Hawaiian Electric Company, the state's investor-owned power utility (IOU), to freeze all new connections. Hawaiians had been installing solar at an unprecedented clip not simply for environmental reasons: forced to generate most of its power with imported oil, the island's electricity rates are an exorbitant three times the national average. As the cost of solar panels dropped significantly, residents of the island turned quickly to renewable power. But the grid wasn't ready for them. Peter Rosegg, a spokesman for the Hawaiian Electric Company, explained the problem this way: "Like every utility in the country, we basically were developed to send electricity one way, from our power plants to our customers. And now we have a situation where 10 percent of our customers are randomly taking or sending power over a system that really wasn't designed for that."[64] The challenge posed by renewable energy to Hawaii's electric system has only grown since 2013: the state now gets 33 percent of its electricity from rooftop solar systems, and the state legislature wants to reach 100 percent renewable energy by 2045.[65] Hawaii is not alone in struggling to incorporate a surge of renewable power into the grid. In January 2018, the US Energy and Information Agency (EIA) announced that almost half of the utility-scale power generation capacity installed in the country involved renewable power.[66] How the grid will handle this surge of renewable energy is far from clear.

The US grid is a ferociously complicated system, by some estimates the largest machine in the world. There are roughly 3,300 electric companies that provide power to individual citizens and commercial users of power in the United States. About two hundred powerful private electric companies—IOUs—are responsible for a significant portion of the net generation (38 percent), transmission (80 percent), and distribution (50 percent) of our electricity.[67] But there are also 2,900 publicly owned utilities and cooperatives that generate, transmit, and distribute electricity; these public entities own and maintain nearly half of the nation's electric distribution lines.[68] In addition, approximately 2,800 independent power producers account for 40 percent of net generation, and the federal government also owns nine power agencies that generate and transmit electricity, including the power produced by large dams like Washington State's Grand Coulee

Dam, facilities that generate 43 percent of the nation's renewable energy. Adding to the complexity of this hybrid system of private and public power companies, the US electric system is made up of more than 7,300 power plants that distribute power through a network of nearly 160,000 miles of high-voltage power lines, linked to millions of low-voltage power lines and distribution transformers that connect hundreds of millions of consumers.[69] Local electric grids are interconnected to form larger networks, all of which are linked together into three main systems: the Eastern Interconnection, which serves the area east of the Rockies and part of northern Texas; the Western Interconnection, which encompasses the states west of the Rockies; and the Electric Reliability Council of Texas (ERCOT), which covers most of the Lone Star State. There are sixty-six balancing authorities spaced out throughout these three main interconnections, whose job it is to make sure that supply and demand are balanced and that the grid continues to function. Balancing must be done in real time: if supply and demand fall out of balance, local, or even large-scale, blackouts across multiple states can result. Most, but not all, balancing authorities are electric utilities or IOUs that have taken on the balancing responsibilities for a specific portion of the power system; sometimes, however, balancing involves siphoning off power from other parts of the grid. Theoretically, the network structure of the interconnections helps maintain the reliability of the overall power system by providing multiple routes for power to flow along and by allowing generators to supply electricity to many load centers: if a power plant or a transmission line goes down in one place, power can be routed to customers from another source. But sometimes the network structure of the electric system can be a problem: in 2003, for example, the nation's second-worst blackout began when a hot spell in the Midwest caused power lines to sag onto some trees that the local power company had failed to trim as it sought to cut costs.[70] The overloaded transmission lines created short circuits that the regional grid operator failed to catch; instead, automatic relays that were programmed to protect equipment by isolating a faulty power line or transformer kicked in one after another, and the crash got bigger and bigger, eventually affecting 45 million people in eight eastern states.

This fantastically complex grid has evolved over the course of more than a century. But today's grid is under unprecedented stress. Blackouts due to extreme weather are increasing, and climate change is only going to intensify this challenge. Even more significant though is the fact that a system designed around large, centralized power plants and one-way flows of power is under strain from new technologies that do not mesh well with this traditional grid architecture.[71] As the troubles of the Hawaiian Electric Company show, renewable energy challenges the one-way model of energy production and distribution that has prevailed since the construction of the modern grid. When people generate their own renewable energy, the grid suddenly needs to be able to cope with a surge of unexpected power. Once enough of this power came online in Hawaii, the integrity of the grid as a whole was threatened—or so the Hawaiian Electric Company claimed—and the utility began to refuse new connections. In addition, since the sun doesn't always shine and the wind doesn't always blow, renewables add lots of what is called "variable power" to the mix, which increases the complexity of balancing the grid. The grid needs to become a lot more flexible in order to cope with these changes. Efforts are underway to boost the system's resilience, but it is not clear that the grid's century-old architecture is ultimately up to the challenge of handling this new complexity.

Most people in developed nations take electricity entirely for granted and know virtually nothing about the challenges faced by the grid. We expect electricity to be there when we flick on the lights or turn on the TV, uninterrupted and stable. It is not that way everywhere: 1.1 billion people—14 percent of the global population—do not have access to electricity today.[72] In many countries, everyday life must be organized around rolling blackouts that are a regular if unpredictable experience. The toll of this grid precarity falls particularly hard on women and children, who do most of the physical work that would otherwise be carried out by electricity. Air pollution from the use of cooking fuel in homes lacking electricity is one of the major causes of respiratory illness and death for young children in low- and middle-income countries.[73] Yet in wealthy countries like the United States, power is both ubiquitous and invisible. Although it is brought to homes by nation- and globe-girdling infrastructures, which continue to hinge on the burning of fossil fuels, electricity seems to arrive out of thin air. It appears clean, helps liberate us from the iron rhythms of day and night, and powers appliances that reduce domestic drudgery. Electricity is a key element of what has been termed the "American way of life," the ensemble of automobiles, highways, suburban single-family houses, and electrified domesticity that spread across the United States after 1945.[74] By powering air-conditioning, electricity has catalyzed a sweeping transformation of the geography and politics of the Unites States, enabling a population boom in the red-state Sun Belt.[75] If electricity does material work, it thus also does political and cultural labor, helping to produce a vision of a world in which individuals imagine themselves as self-determining and autonomous. Electrification has helped power a privatized vision of social space, fueling an entrepreneurial imaginary that has spawned neoliberal policies over the last half century.[76] Ironically, these neoliberal policies of privatization have played a pivotal role in eroding spending on the physical infrastructures of the electric grid.

The precariousness of the grid has been evident for decades. In the wake of the 1977 New York City blackout, which occasioned widespread riots and arson across the city, energy analyst Amory Lovins wrote, "Our electric systems are so brittle because they depend on many large and precise machines rotating in an exact synchrony and the power they generate is delivered through a frail web of aerial arteries which accident, human failure, or human malice can sever."[77] The NYC electric system, Lovins argued, was reliable but not resilient. The grid was built to withstand all but the most freak accidents. Yet, as Lovins rather poetically put it, "possible rare events, each of vanishingly low probability, are infinitely numerous, so we live in a world full of nasty surprises." For Lovins, the lesson of the 1977 blackout for energy policy is "to avoid too much centralization and complexity." Lovins and his

writing partner, Hunter, argued that avoiding catastrophic collapses like the 1977 blackout meant embracing what they called
"soft energy" pathways, which involve shifting from the large,
centralized forms of energy generation and consumption that
have characterized the grid since the early twentieth century to
distributed renewable energy resources and energy efficiency.[78]
Resiliency for the Lovinses is the product of a grid made up of
small, less polluting, relatively self-contained systems. In a world
full of nasty surprises, such systems can be hived off and continue
to operate when other portions of the grid go down, preventing a
total and catastrophic collapse.

Despite the savvy suggestions of the Lovinses on building
grid resiliency, not to mention years of effort to create competitive markets through deregulation, the power system in the
United States today remains highly centralized. Sixty-eight
percent of Americans still get their electricity from IOUs, the
descendants of the monopolistic power companies founded by
tycoons like Samuel Insull and J.P. Morgan. As we will see in the
next chapter, Insull laid out a shrewd blueprint for IOUs that
helped obscure their basic drive to make money: they are hybrid
beasts, regulated by the government and entrusted with delivering power reliably and at an affordable price to the public while
being owned by and charged with paying dividends to investors.
They have also historically been monopolies: when you buy or
rent a house, you automatically become a customer of the local
utility. Deregulation of the energy sector from the 1990s onward
has largely failed since it has not moved many customers off the
historically dominant utilities in particular regions. As a result,

power has remained concentrated in a few hundred IOUs, all of which have large investments in centralized power plants that generate power using fossil fuels. To make matters worse, IOUs have always been interested in encouraging people to consume more energy. This is because they make their money by building grid infrastructure and then charging ratepayers the cost of that infrastructure plus a "reasonable rate of return," as defined by state-level public utility commissions. This imperative to maximize energy consumption and ignore energy efficiency is obviously radically out of line with a world in existential climate crisis. Finally, IOUs follow a paradigm of extraction, situating their dirty fossil fuel infrastructure in or near communities that do not have the political or financial clout to fight them off. For example, coal-fired power plants are most likely to be situated near communities of color and low-income communities; the same thing is true for natural gas infrastructure.[79]

After operating comfortably for over a century, IOUs are now in deep crisis. In 2013, the Edison Electric Institute (EEI), the trade group that represents all US investor-owned electric companies, issued a report that anatomized the threat of what were described as "disruptive challenges" to utilities. Renewable energy, particularly the increasingly affordable combination of solar power and battery storage, is at the heart of these disruptive challenges. When an IOU customer installs solar panels, it hurts the utilities in two ways. First, it diminishes demand for power. In most cases, people with solar panels don't disconnect entirely from the grid since they want access to electricity when the sun isn't shining or when their batteries have run dry. The resulting situation, which is called "partial grid defection," infuriates the utilities because customers continue to use the grid but pay virtually nothing for it since they generate 80 to 90 percent of their own energy. Utilities tend to see these people as cheaters, despite the fact that a fair amount of academic research shows that solar customers tend to save utilities money by cutting demand for energy during peak times of the day like late afternoon, when many people return home from work and switch on their appliances.[80] As customers with solar panels diminish their demand for power provided by the utility by generating their own power, they also cut their electric bills; the utilities then have to charge higher rates to nonsolar customers in order to cover their costs-plus returns. Customers who suddenly see their utility bills spiking for no good reason have a strong incentive to install solar power themselves. The result is an inexorable push toward renewable power that leaves the utilities struggling for income. Following the EEI report of 2013, this dynamic came to be known as the *utility death spiral*.

Utilities in the United States have the stark example of the precipitous decline of big for-profit utilities like RWE and E.ON in Germany as a warning. These companies failed to grasp the threat constituted by renewable energy to their traditional business model of centralized energy generation and distribution. When millions of Germans put up solar arrays on their roofs and formed co-ops to set up wind turbines, creating what is known in Germany as the *energiewende*, or energy transition, the big utilities were caught almost completely off guard.[81] As people generated their own energy, wholesale power prices plunged, eroding the utilities' profits and the value of their power plants by billions of euros. The utilities responded by slashing tens of thousands of jobs. But the hemorrhaging did not stop. Big German utilities are faced with three secular trends that challenge their business models on a basic level: decarbonization, decentralization, and digitization.[82] Like US utilities, German firms like E.ON and RWE have enormous investments in outdated infrastructure, including not just gas-fired but also nuclear power plants, all of which must be retired relatively quickly, given the German government's accelerated timetable for decommissioning nuclear power plants and for decarbonization, which calls for reaching 50 percent renewable power by 2030 and at least 80 percent by mid-century. The shift to decentralized, small-scale forms of energy generation like wind and solar—plus storage, heat pumps, and other new technologies—further challenges the utilities' traditional model of centralized generation. Finally, new digital technologies are knitting small-scale producers and consumers together into what are sometimes called *virtual power plants,* allowing households to trade energy with one another when and as they need it, without the involvement of any central clearinghouse along the lines of the traditional utility. While the destructive economic impact of the sharing economy in sectors like transport suggests that an energy sharing economy might not be an unmitigated boon for consumers, it nonetheless promises to transform the energy industry in the same way that the media business has been upended by platforms like Netflix and Amazon.

Since the death spiral report, the utilities have fought to stymie or at least to slow the adoption of renewable energy in the United States. The Edison Electric Institute itself, for example, has waged a sustained campaign against renewables, donating tens of millions of dollars to sway politicians against renewables and spending an equivalent amount on anti-renewables public relations, while also funding over 150 anti-green nonprofit organizations, including the Koch Brothers–backed American Legislative Exchange Council (ALEC), the US Chamber of Commerce, and the Coal Utilization Research Council dedicated to defeating renewables.[83] These campaigns have centered on spreading fear that the adoption of solar power will inevitably raise rates for nonsolar users. The utilities have also used the capital they derive from ratepayers to support laws and regulations that reduce targets for renewable energy, to campaign for an end to "net metering" laws that force utilities to pay solar customers retail prices for the surplus energy they put back into the grid, and to impose "connection" fees on solar users in order to make up for lost revenues. These measures are not only intended to prevent energy transition but to quash any alternatives to the big utilities. As one director of the Michigan-based renewables advocacy organization Vote Solar put it, the IOUs "want to put a dagger in the heart of rooftop solar."[84] While not all utilities have opposed renewables with the same stridency, the industry as a whole has been on the wrong side of history. Indeed, given the urgency of a transition away from fossil fuels, it is not too much to say that the IOUs have been complicit with the annihilation of the very possibility of history.

THE WILL TO POWER

We live in an age of booming demand for energy in general, and, since fossil fuels continue to supply nearly 80 percent of global energy needs, for dirty energy in particular. The International Energy Agency's (IEA) *World Energy Outlook 2017* predicts a 30 percent expansion in energy demand by 2040, the equivalent of adding another China and India to today's global demand.[85] Although numerous observers project that global coal consumption will flatline by 2040, we should not forget that coal has grown steadily for decades, expanding 741 percent since 1971.[86] The IEA says that China's coal consumption will decline by 15 percent by 2040, but coal capacity there grew by a factor of five from 2000 to 2017, to reach half the world's total.[87] As its economy has soured in recent years, China has begun bringing more coal on steam again, and there is also evidence that China is now exporting significant amounts of heavy industry and, with it, coal-fired power plants to other Asian nations.[88] Demand for fossil energy is up significantly in India, Turkey, Indonesia, and other developing nations where key decisions about twenty-first-century infrastructure are being made.[89] In addition, the IEA projects that global demand for oil will continue to grow, if at a decreasing pace, and that fossil gas use will rise by 45 percent by 2040. The IEA's bullish projections for fossil fuels are probably influenced by the fact that at least two of the authors of the *World Energy Outlook* were staff on leave from oil companies, which continued to pay their salaries while they wrote the projection document.[90] These projections certainly square with those of Big Oil companies like ExxonMobil and Shell, which both anticipate continuing production of fossil fuels well beyond mid-century. Their rosy outlook is no doubt supported by the fact that, despite pledging in 2009 to end subsidies for "inefficient" fossil fuels, rich countries have set no deadline for stopping such support, and fossil fuels of all kinds continue to receive hundreds of billions of dollars a year in public funding.[91]

All of this is radically out of line with any scenario that averts a hurtling course to fiery apocalypse. Indeed, according to recent research, current climate policies have us headed for at least 3° of warming.[92] To avert this catastrophic future (which is already obliterating the present for many parts of the planet), we need a virtually immediate cessation of all fossil capitalist activity. Analysis by Oil Change International and a coalition of allied environmental organizations has shown that already-developed fields and mines contain enough coal, oil, and gas to take the world beyond 2°C.[93] Reserves in currently operating oil and gas fields alone, even with no coal, would take the world beyond the 1.5° goal that policymakers announced as the aspirational goal and safe limit in the Paris Agreement. As this analysis shows, additional investment in fossil fuels would either leave those assets stranded or would drive the climate toward apocalyptic upheavals. Although planetary ecocide is obviously far and away the greatest danger, the prospect of a global economic collapse unleashed by trillions of dollars of worthless investments in fossil fuels that cannot be extracted is highly alarming as well, particularly since the burden of this massive liquidation of capital would likely fall on the most vulnerable nations and communities rather than simply on the coal and oil barons. This is why Oil Change International and its allies call for a managed shift away from fossil fuels. This transition should be led by countries with developed energy sectors, which support other countries on a path to fossil-free development as a form of reparation for their colonization of the atmospheric commons. And this shift must be rapid. Even conservative estimates suggest that greenhouse gas emissions in rich nations like the United States must begin rapidly declining by 2020. Time is not running out—it's up!

Given this fact, why is there not more panic in official circles about the relatively slow rate of the energy transition? The lack of a requisite sense of urgency among policymakers is linked to the wishful thinking baked into dominant scenarios for mitigating climate change. Most models in circulation today, which are framed around giving us a 66 percent chance of meeting the Paris Agreement's 2°C warming target, assume that greenhouse gas emissions will be net-zero by mid-century. But they also predict that fossil fuel companies will not only continue to produce but will even keep exploring and extracting until 2050.[94] The scenarios for the future advanced by the IEA, the Intergovernmental Panel on Climate Change, and an oil company like Shell are thus fundamentally similar.[95] As Shell puts it in their *Sky Scenario*, "From 2018 to around 2030, there is clear recognition that the potential for dramatic short-term change in the energy system is limited, given the installed base of capital."[96] In other words, Shell and other corporations and global elites simply have too much money sunk in existing fossil capital to immediately shut down fossil fuel extraction. To prevent a serious disruption to the global financial system—and, it need hardly be mentioned, the fossil fuel industry—we need to wait for all the existing wells to run dry, mines to shut down, power plants to close, factories to retrofit, and new technologies to scale up. This will take several decades, hence the plans to keep on with business as usual until mid-century.

But how can fossil fuels be part of an energy system that produces no carbon in 2050? The answer lies in the "net" part of net-zero emissions: all of the dominant scenarios for mitigation assume that continuing greenhouse gas emissions will ultimately be pulled out of the atmosphere and sequestered in the earth, leading to "negative emissions."[97] We will be able to keep pumping carbon into the atmosphere until 2050, in other words, driving temperatures above 2° in what is referred to as "overshoot." But then in the second half of the century, we will suck that carbon back out of the atmosphere in quantities vast enough to restore the balance and pull temperatures back below 2°. But this view, accepted by most official policy bodies the world over, is an example of the most outrageously reckless, swivel-eyed wish fulfillment that one can image. As Kevin Anderson of the Tyndall Centre for Climate Change points out, we talk about negative emissions technologies today as if they actually exist, but they don't—"they're just sketchbooks on computers."[98] So-called Direct Air Capture technologies may prove to be part of a future clean energy technology mix, but they are not now and they are never likely to be the silver bullet for climate change many people imagine them to be because they are not competitive economically with existing renewable energy technology, and the amounts of carbon that would have to be sequestered to produce negative emissions are truly mammoth.[99] Shell's *Sky Scenario*, for instance, would require up to eleven gigatons of carbon dioxide to be buried each year by 2070, an amount equivalent to one-third of today's global emissions.[100] In addition, the 2°C threshold that forms the baseline for all these mainstream mitigation scenarios is not a safe threshold, as policymakers themselves indicated in the Paris Agreement. Limiting warming to 1.5°C means reaching negative emissions a decade earlier than most of these scenarios anticipate.[101] In addition, the idea that we will be able to overshoot—heating the planetary system to 2°C or more and then magically reversing this warming—ignores the potential tipping points and feedback effects that would be unleashed by such dangerous warming. After all, global warming is not simply produced by the release of greenhouse gases resulting from various human activities, but also by the diminishing of the albedo effect as the polar icecaps melt, by the release of methane from reservoirs beneath the Arctic tundra, and the drying out of rainforests, among other forms of climate change–induced environmental feedback.[102] In sum, dominant scenarios for mitigating climate change are based on wishful thinking in which the unproven technologies of the future prevent us from having to engage in radical emissions cuts in the present.

In the face of these alarming realities, mainstream commentators tend to speak in hyperbolic terms about the increasing economic competitiveness of renewable energy. It is undeniably true that renewable energy has been making remarkable strides in recent years. According to REN21, the global renewable energy policy network, 2017 was a record-breaking year for so-called modern renewables like wind and solar power, seeing the largest-ever increase in renewable power capacity, falling costs, and significant increases in investment.[103] REN21 notes that renewables accounted for an estimated 70 percent of net additions to global power capacity in 2017, "due in large part to continued improvements in the cost competitiveness of solar PV and wind power."[104] Meanwhile, REN21 notes, the battery industry is coming into its own, making storage of variable energy sources like solar a possibility. REN21's analysis is a fairly common one: the falling cost of renewables is often hailed as particularly significant since it is seen as making renewable energy competitive with or even cheaper than fossil fuels, thereby allowing the market to do its magic of transforming the energy sector without any overt political intervention. The decreasing cost of renewables leads to many breathless headlines that suggest that we are winning the battle for a transition to green power. *Renewable Energy Focus* offers a typical example: "Onshore wind power now as affordable as any other source, solar to halve by 2020."[105]

The sense that we are undergoing a rapid transition as a result of the growth of renewables is affirmed by projections for the future. A recent study published in *Nature Climate Change* for instance describes a "carbon bubble." Current investment in fossil fuels, the study concludes, will become stranded "irrespective of whether or not new climate policies are adopted" as a result of "an already ongoing technological trajectory."[106] It is argued that the carbon bubble will burst because of advances in technologies for energy efficiency and renewable energy, developments that are already making low-carbon energy more economically and technically attractive and that will generate an inevitable, rapid decline in demand for fossil fuels. If the nations of the world embrace aggressive carbon mitigation goals, the study suggests, the carbon bubble is likely to burst even sooner. But despite his own study's optimistic projections for a transition away from fossil fuels, Jean-François Mercure, the study's lead author, cautions that the transition is happening too slowly to stave off the worst effects of climate change.[107] To keep within 2°C above preindustrial levels, much stronger government action will be required, Mercure warns.

Yet even if technological innovations and market forces are pointing toward a gradual shift away from fossil fuels, in the long run, as the economist John Maynard Keynes once quipped, we are all dead. Renewable energy is simply not growing fast enough to shift us away from our current trajectory toward planetary ecocide. Sean Sweeney and John Treat of Trade Unions for Energy Democracy (TUED) have long warned that supporters of a market-led transition trumpet an optimistic narrative that obscures key contextual details about the energy transition.[108] In a series of essential working papers published for TUED, Sweeney and Treat point out that green capitalist narratives about energy transition are selective, both in terms of their time frame and their geography. If we expand analysis beyond the last five years or so, the peril of this sanguine but blinkered outlook becomes clear. The *Statistical Review of World Energy* report by British Petroleum, for example, an assessment that many analysts use to understand the evolution of the energy sector on a global level, concludes that fossil fuels were responsible for 38 percent of generation in the electricity sector in 2017—*exactly the same proportion as in 1998*.[109] As Spencer Dale, BP's chief economist, put it, "there has been almost no improvement in the power sector fuel mix over the past 20 years [. . .] this is one area where at the global level we haven't even taken one step forward, we have stood still: perfectly still for the past 20 years."[110] Renewables have certainly increased in recent years, in other words, but at the same time fossil fuels—particularly fracked oil and gas—have increased and nuclear power has decreased. Meanwhile, coal had a tremendous growth spurt beginning in the early 2000s, especially in developing countries like China and India. Even after the central government tried to rein in this coal surge in China, local authorities kept building coal-fired power plants in defiance of Beijing.[111] Now, the Chinese government has begun exporting coal power through its Belt and Road Initiative to developing countries like Bangladesh, Kenya, and Vietnam, where demand for power is swelling.[112] As a result, the overall proportion of global electricity coming from nonrenewable sources has remained *roughly the same* over the last two decades.[113] But even this dismal assessment is too upbeat since BP's chief economist fails to note that since electricity consumption is far higher today than it was in 1998, in absolute terms the 38 percent of present fossil fuel consumption is generating emissions markedly greater than the same proportion was at the time of the signing of the Kyoto Protocol.

BP's gloomy analysis is not unique. According to the recent reports from the International Energy Agency (IEA), "the world is currently not on track to meet the main energy-related components of the Sustainable Development Goals, agreed by 193 countries in 2015."[114] The IEA has been accused of low-balling its projections for the growth of wind and solar power, but it is not alone in its restrained assessment of the growth of renewable energy.[115] The United Nations echoes this conclusion, commenting that, as of 2017, "progress in every area of sustainable energy falls short of what is needed to achieve energy access for all and to meet targets for renewable energy and energy efficiency. Meaningful improvements will require higher levels of financing and bolder policy commitments, together with the willingness of countries to embrace new technologies on a much wider scale."[116] Exactly how much progress is necessary to avoid climate catastrophe? The International Renewable Energy Association (IRENA), following a forum in mid-2018, writes that "increasing the speed of global renewable energy adoption *by at least a factor of six*—critical to meeting energy-related emission reduction needs of the Paris Climate Agreement—can limit global temperature rise to 2°."[117] Given that one of the signal achievements of the Paris Agreement was the recognition by global leaders that 2°C is not a safe state for the planet, and that warming should be no more than 1.5°, we can only assume that IRENA's goal of a sixfold increase in renewable energy adoption is far too low to avoid climate chaos. Yet, according to the global renewable energy policy network REN21, the average growth rate of modern renewables over the last decade was just 5.4 percent.[118] This helps explain REN21's sober analysis of the growth of renewable power in its 2018 *State of the Global Renewable Energy Transition* report. While noting that "renewable electricity is now less expensive than newly installed fossil and nuclear energy generation in many parts of the world . . . and less expensive even than operating existing conventional power plants in some places," REN21 states that "the global energy transition is only fully underway in the power sector; for other sectors it has barely begun. The power sector on its own will not deliver the emissions reductions demanded by the Paris climate agreement or the aspirations of Sustainable Development Goal 7 to ensure access to affordable, reliable, sustainable and modern energy for all."[119] While the share of modern renewable energy in the total global energy supply is on the rise, "the heating and cooling and transport sectors, which together account for 80 percent of global total final energy demand, are lagging behind," according to the REN21 report.[120]

But what of the widely accepted idea that economic growth has become "decoupled" from emissions, an idea that was the linchpin of President Barack Obama's conclusion in his much-cited 2017 editorial in the journal *Science* that the momentum of the energy transition was "irreversible"?[121] In recent years, many pundits have joined Obama in celebrating the purported flat-lining of emissions rates, and it was assumed that the world had turned a corner to clean growth. The World Resources Institute, for example, reported in 2016 that twenty-one countries had successfully decoupled growth and emissions.[122] This idea of disconnecting disastrous greenhouse gas emissions and economic growth, which is also known as *dematerialization,* became a key doctrine of green capitalism. The dogma of dematerialization has inspired recent documents such as the *Ecomodernist Manifesto* (2015), whose authors proclaim with no apparent sense of irony that "intensifying many human activities—particularly farming, energy extraction, forestry, and settlement— so that they use less land and interfere less with the natural world is the key to decoupling human development from environmental impacts."[123] But the cardinal early articulation of the idea of decoupling is the *Stern Review* (2006), in which the former chief economist for the World Bank set the dominant outlook for global efforts to address the climate crisis for years to come. Despite sounding a critical note about the functioning of the capitalist system, Stern nonetheless went on to assert that "the world does not have to choose between averting climate change and promoting growth and development. Changes in energy technologies and in the structure of economies have created opportunities to decouple growth from greenhouse gas emissions."[124] You can have your capitalist cake and eat it too, in other words.

The primary solution advocated by Stern in his influential report was emissions trading, also known as cap-and-trade. Yet the most prominent example of such cap-and-trade policies— the European Union Emissions Trading Scheme (EU ETS)— utterly failed to diminish carbon emissions. Established in 2005, the EU ETS allocated emissions to industries like power plants and manufacturing plants, allowing those that emitted less than their quota to sell rights to pollute to other industries. But EU member states with a vested interest in protecting parts of their energy sectors overallocated allowances, allowing polluting industries to keep polluting *and* also to make windfall profits by selling their unused allowances.[125] Offsetting projects that went along with emissions trading such as monocultural timber plantations sited in poor nations had devastating effects on local forest-dwelling communities.[126] During the initial period of the ETS, the price of carbon fell to zero as a result of over-allocation. As a form of "privatization of the air," emissions trading helped fuel the rise of financial speculation on emissions.[127] The United States tried to develop its own version of emissions trading through the Climate Action Partnership, which was a collaboration between centrist environmental organizations like the Environmental Defense Fund and the NRDC and polluting industries like General Electric, BP, and Duke Energy. This effort eventually led to the American Clean Energy and Security Act of 2009, a bill that died an ignominious death in Congress after it became clear that it would suffer from the same corruption as the EU ETS, and would also gut the ability of the EPA to regulate carbon emissions from power plants. A version of the Climate Action Partnership was recently resurrected with a proposal from the immodestly named Climate Leadership Council for a carbon tax and dividend scheme. Backed by a coalition of Bush-era lobbyists like Trent Lott and Big Oil companies like ExxonMobil and BP, the carbon tax plan imposes the mildest possible carbon tax on polluters, with unspecified future increases, in exchange for revoking EPA regulations meant to curb greenhouse gas emissions and giving immunity to energy companies from climate-based lawsuits.[128] It should not be surprising that Big Oil is behind this carbon tax proposal: like emissions trading, this scheme poses no firm limit, or "cap," on growth. Fossil capital has understood that it needs to be seen to be doing something about planetary ecocide, but it fully intends to continue with business as usual.

Celebrations of the delinking of economic growth and carbon emissions help to perpetuate such disastrous policies. But the dematerialization of the economy promoted by green capitalists in recent years is nothing but a sophisticated falsehood. Growth and emission increases have not been decoupled, at least not for more than a temporary period. The global economy has been heating back up in recent years: energy demand grew at a rate of 2.1 percent in 2017, which is twice the average rate of increase during the previous five years.[129] This is *very* bad news for carbon emissions: indeed, according to REN21, carbon emissions from energy consumption increased by 1.4 percent in 2017, after holding steady during the previous four years.[130] And even if the trend toward flatlining of emissions rates had persisted, this trend was based on a highly circumscribed geographical analysis of the relation between economic growth and greenhouse gas emissions. After all, the countries celebrated as having successfully decoupled by organizations such as the World Resources Institute have all shifted most of their heavy industrial production to countries like China and Mexico in search of cheaper labor over the last four decades. Indeed, if carbon embedded in products imported from low wage countries were taken into account (as what is known as "consumption-based accounting" seeks to do), emissions rates in wealthy countries would show a significant *increase* over the last half decade.[131] Statistics about diminishing rates of emissions in rich countries thus tell us more about the accumulation strategies of global elites through the construction of a new international division of labor than they do about the sustainability of capitalism as a world system. Add to this the fact that diminishing *rates* of emissions are not the same as diminishing emissions. In 2013, the Mauna Loa Observatory in Hawaii recorded its first-ever carbon dioxide reading in excess of four hundred parts per million.[132] Just four years later, that figure is above 415ppm and rates of emissions are escalating sharply. Even if emissions rates were to be successfully cut for an extended period rather than for a relatively brief interregnum of economic downturn in the growth-oriented capitalist system, it would be wholly inadequate to forestall environmental cataclysm. Only when emissions *are cut in half* will atmospheric carbon dioxide— which stays in the atmosphere for centuries—begin to level off.[133] But this is not going to happen, since global elites understand that cutting emissions seriously would imply cutting growth. And cutting growth is anathema to the capitalist system.

Indeed, despite their pledges in the Paris Agreement to slash carbon emissions dramatically, rich countries continue to endorse expansion of fossil fuel production. At a meeting in mid-2018, energy ministers from the world's biggest 20 econo- mies (aka the G20) released a communiqué deploring the Trump administration's attacks on the Paris Agreement, but at the same time committing their countries to "expand [fossil gas produc- tion] significantly over the coming decades."[134] These two pro- nouncements are directly contradictory. Fossil gas, often referred to as "natural gas," the benign-sounding name preferred by the industry advocates and government officials, is often described as a "bridge fuel" in the transition toward lower emission energy systems. The continuing circulation and wide acceptance of this myth is, as Bill McKibben argues, the greatest failure of the cli- mate movement.[135] The myth of fossil gas as a bridge fuel has pen- etrated elite opinion to such an extent that the State Department even set up an agency to export fracking technology to other countries under President Obama.[136] But the main component of "natural" or fossil gas is methane, a gas that warns the atmosphere *eighty-six times as much as* carbon dioxide.[137] This means that even small leaks during the extraction of fossil gas can have a major impact on climate change. In fact, most studies suggest that fracking in the United States is characterized by leakage rates of 3 percent or higher.[138] So, although the United States may have shuttered a significant number of coal-fired power plants in recent decades, the massive expansion of fracking in recent years means that, as Bill McKibben puts it, "[We're] still pouring greenhouse gases into the atmosphere at pretty much the same rate as before."[139] When it comes to climate change, standing still on carbon emissions rates is actually running full tilt over a cliff toward ecocide.

THE FINANCIALIZATION OF FOSSIL CAPITAL

Kip "The Whip" Oliphant, the flamboyant CEO of Dark Elephant Energy, a fracking company operating in some of the most profitable shale plays in the United States, is an outrageous amalgam of traditional Texas oilman and New Age guru. Kip—one of the most amusing characters in Jennifer Haigh's *Heat and Light*, a novel exploring the cultural, economic, and political ramifications of fracking—spouts self-help aphorisms to his company's good-ole-boy shareholders as he attempts to convince them to back his company's expansion efforts in the Marcellus Shale in rural western Pennsylvania: "Now is the time to leverage our first-mover advantage," he tells them. "We're on the verge of a new inflection point."[140] The success of Dark Elephant derives from a "Texarkana crackpot named Wade Dobie," who joined the company's engineering team over a decade before the fracking boom and sold Kip on the horizontal drilling technology that is one of the core elements of fracking, an extreme form of extraction that involves injecting undisclosed explosive chemicals into the ground to pulverize rock formations that trap oil and gas—chemicals that often wind up in people's drinking water.[141] Through Kip, the novelist gives readers a sense of the inscrutable forces—both geologically subterranean and politically obscure—that shape the lives of the myriad characters who converge in the town of Bakerton, Pennsylvania. After his platitude-studded speech to shareholders near the outset of the novel, Kip almost entirely disappears from the plot, which focuses on the constellation of local residents and roughnecks who populate Bakerton during the fracking boom. Yet if *Heat and Light* shares a collective approach to character and form with other novels of extraction such as Emile Zola's *Germinal* and Abdelrahman Munif's *Cities of Salt*, unlike these works Haigh's fiction is not a celebration of the collective capacity of worker power and popular resistance to subvert and transform fossil capitalism.[142] A product of the age of neoliberal globalization, *Heat and Light* is a heartbreakingly disillusioned novel whose chief embodiment of anti-fracking activism, a charismatic geology professor named Lorne Trexler, is just as unscrupulous in his manipulation of Bakerton residents as are the representatives of Dark Elephant, who rove the countryside trying to convince locals to sign leases on their property for a pittance.

Appropriately enough for this cynical novel of a dispirited age, it is not the rabble-rousing efforts of Trexler or the other characters he helps educate and mobilize that ultimately eject Dark Elephant from Bakerton. Instead, Kip is brought down by market dynamics. As an oily TV anchor named Ty Slater informs us late in the novel, "Boys and girls, energy is and forever will be a numbers game. Right now the numbers don't work. They don't even *almost* work. The past few years we've seen a massive rush to drill, and now we're *drowning* in natural gas. The market is so glutted they're practically giving the stuff away. Trouble is, drilling for gas in these particular formations is hella expensive. With gas at a buck ninety, they can't cover their operating costs."[143] Kip, Slater observes, is an addict. But unlike many of the residents of Bakerton who are ravaged by the opioid crisis and the deeper emotional malaise that spawns it, Kip is "addicted to drilling."[144] Kip's contract makes him part owner of every well the company drills, as long as he covers his share of the drilling cost. But with massive overproduction of fossil gas from the Marcellus shale, Kip is deeply in debt, and has cashed in Dark Elephant's stock in order to keep drilling. When sinking share prices eventually catch up with Kip's smooth talk and shady financial dealings, the good-ole-boy shareholders give Kip the boot. Rather than any popular groundswell against fossil capitalism, in other words, it is the peculiar technical and financial characteristics of fracking that doom Kip and his company's exploitation of Bakerton.

Heat and Light was published in 2016, a moment when a financial meltdown of the fracking industry seemed imminent.

On March 2 of that year, an SUV driven by Aubrey McClendon, billionaire CEO and co-founder of fracking giant Chesapeake Energy Corporation, careened out of its lane and slammed into a concrete wall at seventy-eight miles an hour. The day before his death McClendon had been indicted by a federal grand jury on charges of rigging bids with competitor companies in order to suppress the value of land where the companies intended to drill. McClendon's indictment threatened to shed harsh light on the fracking industry's unscrupulous business dealings, as well as on his own increasingly indebted state.[145] Although investigations eventually failed to conclude that McClendon's death was a suicide, his smashup functioned as a potent metaphor for the state of the industry he seemed in so many ways to embody. As Haigh's fictional TV anchor observes, the fracking industry seemed to be consuming itself in a binge of overproduction.

Although the technologies behind fracking were invented in the early 2000s, fracking really took off after 2008. Between 2010 and 2015, US oil production exploded as a result of fracking, growing from 5.4 million barrels per day to nearly 10 million barrels daily five years later.[146] This unparalleled surge in production caused global crude markets to plummet in late 2014. Particularly significant was the refusal of Saudi Arabia to cut production in order to prop up prices, a move that many took as a reflection of the Saudis' desire to smash the US fracking industry.[147] Sure enough, by 2015, many fracking companies could no longer afford to pay their creditors; declarations of bankruptcy in the industry left lenders and investors more than $70 billion in the red.[148]

But the fracking industry survived this crisis. To understand how it was able to do so, we need to go back to the moment when the industry matured. Before 2008, despite the development of the combination of technologies that facilitated fracking earlier in the decade, US fossil fuel production was in a state of lengthy decline. Domestic oil and gas wells had stagnated for much of the early 2000s, with production flatlining at around five million barrels.[149] Industry experts warned that fossil fuel shortages could be imminent, sparking jeremiads such as James Howard Kunstler's *The Long Emergency*.[150] As oil supplies peaked, commentators such as Kunstler argued, scarcity of energy and food would lead to mass starvation and collapse of the United States' unsustainable suburban geography. These apocalyptic pronouncements turned out to be dead wrong: the United States was in fact on the verge of one of the greatest expansions of fossil fuel production in history. In late 2007, a group of oil industry veterans met at a conference organized by the bank Goldman Sachs to sell investors on the profits to be made from the new technology of hydrofracking. Their timing was perfect. At this point, fracking companies like EOG Resources (a company established in the wake of Enron's collapse) were on shaky financial ground since they had spent hundreds of millions of dollars trying to make fracking work. Fracking might have been abandoned to the dustbin of history, but less than a year later, banks like Goldman Sachs found themselves suddenly awash with taxpayer money after the Federal Reserve decided to bail them out following the financial crash of autumn 2008. The banks took the public money forked over to them by the Troubled Asset

Relief Program (TARP) and plowed it straight into fracking. Bailout recipients like J.P. Morgan Chase and Citigroup loaned over $900 million and $600 million respectively to EOG alone.[151] With oil prices spiking, fracking looked like a good investment to the banks. The Fed's low interest rates, intended to help the housing market rebound, propped up fracking further by allowing fracking companies to borrow trillions of dollars from banks at negligible interest rates.[152]

How did the fracking industry survive the crisis of overproduction that destroyed figures like Kip Oliphant and Aubrey McClendon? As we have seen, the notion that contemporary fossil capitalism in the United States is a free market is a pernicious illusion, one that obscures the role of the state in propping up an industry that is unsustainable in every sense of the term. The low interest rates maintained by the Federal Reserve after the 2008 financial crisis meant that fracking companies could refinance their debt even as their losses mounted when overproduction drove oil prices down. The industry thus survived on a lifeline established by the US state. Under the Obama administration, the Federal Reserve essentially bankrolled the development of a whole new regime of fossil fuel capital. In 2000, only 2 percent of oil and gas wells drilled in the United States were for fracking; today, 69 percent of wells are fracked.[153] The United States now produces more fracked oil than its entire oil output a decade ago. The result has been an environmental catastrophe. As fracking more than doubled oil production, driving prices down, consumption went up. In 2015, Obama lifted the longstanding ban on oil exports. The United States, now producing more oil and gas than both Saudi

Arabia and Russia, is building pipelines and other fossil fuel infra-structure helter-skelter with the aim of exporting its excessive production to other countries. The massive public subsidies that spawned the fracking boom could obviously have gone instead to support the growth of renewable energy, which fossil capitalist minions ironically like to deride as excessively dependent on pub-lic support. But instead the US state chose to prop up the frack-ing industry in the face of the clearly apocalyptic environmental implications of such idiotic economic investments.

But aside from its dire environmental implications, frack-ing is not even a good bet in economic terms. Since 2007, energy companies have lost an estimated $280 billion on shale invest-ments.[154] These astronomical losses stem in part from the fact that the output of fracking wells declines much more rapidly than traditional oil wells. In North Dakota's Bakken formation, for example, output from the average well decreases over 85 percent during the first three years of production. As a result, thousands of fracked oil wells are producing much less than their own-ers forecast to investors.[155] Analysts of the industry are already beginning to fret about "peak shale."[156] Indeed, only five of the top twenty fracking companies actually made money in 2018.[157] But Wall Street bankers and hedge fund managers don't seem to care whether the companies they are lending to and investing in actually make money: they are getting rich making loans that put the industry ever deeper in debt.[158] Fracking is thus a blatant con game, an industry based on a publicly supported Ponzi scheme in which unprofitable companies burrow ever deeper into unsus-tainable debt to produce planet-destroying extreme energy. The

New Age rhetoric of Kip "The Whip" Oliphant in the novel *Heat and Light* is a remarkably accurate account of the hyperbolic con game necessary to keep this industry afloat.

Such a Ponzi scheme must be constantly fed by fresh infusions of taxpayer money. The latest instance of this comes in the form of the Trump Tax Cuts and Job Act of 2017. EOG Energy, routinely touted as one of the most solid shale oil-and-gas companies, lost $1.1 billion in 2016; the following year, after the Trump tax cuts, EOG reported a net income of $2.6 billion—85 percent of which was due to the new tax law.[159] Much of Trump's tax cuts went to wipe away the debts acquired by fracking companies over the last decade. Big Oil companies like ExxonMobil have promised to use the tax cut windfall to expand their production, which is likely to lower the price of oil further, putting yet more pressure on highly leveraged shale companies. The economic contradictions and pressure will thus mount ineluctably higher. As one critic of the industry puts it, "Just as most sharks must swim to stay alive, shale companies must drill to preserve CEO bonuses, which are often tied to oil production, not profits. So, they drill. Even when that means losing money on nearly every barrel of oil they pump."[160] It is on this unsustainable foundation that the fracking boom is based. But fracking is not just a planet destroying energy regime: it also underpins the rhetoric of "energy independence" and *Make America Great Again* that Trump uses to justify his extractivist populism. As the financial and technical contradictions mount, this rotten state of affairs and the corrupt regime of extraction that it is based on teeter ever closer to collapse.

CONCLUSION

The previous sections of this chapter have shown that there are three key, interwoven sectors that must be tackled and transformed if energy transition is to happen with the speed and democratic accountability the planet needs: 1) fossil capitalism, 2) finance, and 3) state power. The next chapter deals in detail with the question of the state. I conclude this chapter with some suggestions about how the former two sectors might be redirected to work for the public and planetary good. As we have seen, the free market is an illusion, and the idea that market forces will bring about a quick and just energy transition by themselves is folly. Under the capitalist economic system, the lion's share of capital allocation is determined through the private banking and financial system. The goal of the system as it currently stands is to maximize profits for a small number of hyper-wealthy investors rather than to benefit society as a whole. Over the last four decades, as this system has been deregulated and increasingly integrated globally, staggering economic inequalities have opened up around the world and economic volatility has increased dramatically. Financialization has also intensified the ecological crisis since financiers prefer the high short-term returns to be gained from ecosystem-annihilating fossil fuel projects to investments in renewable energy and green infrastructure. With its emphasis on exponential growth over a short time scale, the capitalist system is now driving planetary ecosystems toward catastrophic crisis. But if we are to move beyond simply decrying the insidious impact of free market fundamentalism and ecocidal capital, what alternatives are needed in the short term?

We have seen that financialization in general and the Federal Reserve in particular have played a pivotal role in the fracking revolution. It follows that these pillars of contemporary capitalism either need to be reformed or abolished. If the latter course is adopted, alternatives must be created. In their tightly argued discussion of how to assert public control over the financial system, Johanna Bozuwa and Thomas Hanna of the Democracy Collaborative suggest that socialization of this sector requires two intersecting interventions: democratization of the Federal Reserve, and a large-scale expansion of public and cooperative banking.[161] The Federal Reserve is essentially the nation's central bank; like similar institutions in other countries, it has the ability to print money, set interest rates, and dictate monetary policy. But the Fed is structured differently from other central banks. The nation's twelve regional Federal Reserve banks are set up like private corporations: large private commercial banks hold stock in these Federal Reserve banks and earn dividends on that stock. They also elect the majority of each regional Federal Reserve Bank's Board of Directors. Based in Washington, DC, the Fed's central Board of Governors is politically appointed—by the sitting US president and the Senate—but has traditionally been fiercely independent and hostile to all efforts at government oversight. While such autonomy is usually represented in the news media as a positive thing, as insulation from the whims of particular politicians, in effect it means that the interests of finance capital almost exclusively animate the Fed.[162] The damaging impact of the Fed's structure and orientation on the populace in general was made abundantly clear during the Great Recession, when the Fed showered seven hundred billion dollars on the banking industry while abandoning millions of Americans to foreclosure, bankruptcy, unemployment, and homelessness. As we have seen, the Fed's post-Recession policies also kick-started and sustained the fracking revolution, one of the most momentous and destructive environmental developments in the nation's—and the planet's—history.

Like the nation's IOUs, the Federal Reserve is essentially a private organization disguised as an institution serving the public interest. A genuinely publicly owned and democratically accountable central bank, as Fed expert William Greider puts it, would understand that "its obligation is to society, not money markets."[163] Congress created the Fed, and Congress can transform this antiquated and grossly biased institution whenever it wishes. Among the reforms suggested by Bozuwa and Hanna are a new structure that drastically reduces the power of commercial banks, a new charter that unequivocally prioritizes social benefits and ecological sustainability, heightened transparency and accountability standards, greater integration with other fiscal policymakers in the government, and robust mechanisms for stakeholder and community participation in decision-making processes. With such reforms in place, the Fed could play a major role in addressing climate change, ensuring that capital is invested in renewable energy in particular and in all of the related sectors of US society that need to be transformed as part of a Green New Deal.

But the reformed Fed cannot and need not do this alone. A significant expansion of public and cooperative banking must also play an important role in providing financing for energy transition and related green infrastructure. Despite years of neo-liberal privatization, public banks account for a stunning quarter of all banking assets, and are major supporters of infrastructure and development around the world.[164] Since public banks are not burdened with generating high rates of return for rich investors, they are able to invest in vital public services and crucial development needs, and are also able to provide lower costs for consumers and increased economic stability to boot. Public banks are certainly not perfect, and in the current neoliberal conjuncture they are under pressure from private investors to partner in infrastructure projects, thereby providing public guarantees that socialize the risks of green infrastructure while facilitating the privatization of returns.[165] But public banks can potentially be used to finance a massive build out of renewable energy infrastructure, public transportation, green public housing, and other key elements of a just transition. As with the reformation of the Fed, to function effectively public financing must be conceived of as a common good and the governance structures of public banks must be thoroughly democratized. A combination of state-owned and alternative cooperative and worker-managed banks is perhaps the best combination to transform the financial sector from the roots up and to provide the much-needed infusion of investment for energy transition.

Yet it will take more than just large-scale finance for energy transition to avert planetary ecocide. In addition, the obstructionist investor-owned utilities must be sidelined or dismantled. To speed the transition to renewable energy, Bozuwa and Hanna of the Democracy Collaborative have proposed the creation of a Community Ownership of Power Administration (COPA).[166] Modeled on the New Deal's Rural Electrification Administration, COPA would address the gross inequities perpetuated by fossil capitalism against poor communities and people of color by funding energy transition initiatives that also build community wealth; examples of such uplift include "localized procurement processes, robust workforce development for a just transition, and worker-centered labor agreements."[167] Wind turbines, for example, could and should be manufactured and assembled in historically disadvantaged communities of color in order to redress economic marginalization and environmental injustice. Such local procurement strategies would have the additional crucial benefit of ensuring that such equipment does not have to be transported great distances before being installed—with all the carbon emissions that such transport would entail. In order to guarantee that local communities have adequate democratic oversight of projects related to the energy transition, Bozuwa and Hanna recommend the creation of various institutions for community empowerment, from neighborhood assemblies, to multi-stakeholder boards, to participatory planning/budgeting processes. This kind of grassroots input into the energy transition would prevent it becoming inflexible, bureaucratic, and unresponsive to local needs—as the IOUs currently are. With the significant infusions of capital that a system of public and cooperative banks and a democratized central bank could provide, COPA would be empowered to support energy transition projects on multiple scales, from local solar cooperatives to regional public power authorities along the lines of the Tennessee Valley

Authority (which is itself a product of the struggle for pub-lic-sponsored electrification during the New Deal).[168] Support for energy transition on such differentiated scales would help disag-gregate the current monolithic and brittle grid, generating a pro-fusion of local initiatives that would be well positioned to nimbly embrace the revolutionary trends unleashed by the three "Ds"—decarbonization, decentralization, and digitalization—while operating for the benefit of the public in general rather than a narrow stratum of wealthy investors. COPA would help sideline the obstructive IOUs while ensuring that their unionized labor forces were offered comparable well-paying jobs in the renewa-ble energy sector.

Of course, IOUs are not the only barriers in the way of energy transition. Fossil capitalism in general needs to be dismantled. To pull us out of the fossil capitalist death spiral, the fossil fuel indus-try needs to be nationalized and wound down with the great-est possible speed. Other proposals such as carbon taxes have proven toothless at the low levels at which they have been imple-mented. To function as intended, they would need, as the IPCC has suggested, to be set at levels so astronomical as to effectively ban fossil capitalism. But this supply-side solution would not give the public any democratic say over how the fossil fuel industry was shut down. Given the short amount of time and the danger of precipitating an economic crash that would devastate the lives of ordinary people, a carefully planned program to end the life of these climate change–denying corporations is necessary.[169] A convincing program for the dismantling of fossil capitalism might be based on the policies the Federal Reserve adopted in the wake of the financial crash of 2008.[170] The government would slash current massive subsidies to fossil capitalism and enact robust regulations that would cause the value of fossil fuel corporations to drop. The state would then buy out these corporations at a relatively low cost, one that reflects the unburnable reserves on their books as well as their historical culpability for ecocide. In order to avoid the moral hazard implicit in government ownership of still-lucrative but doomed industries, explicit plans would be articulated for the speedy dismantling of fossil capitalism: all new exploration would be banned, the moratorium on fossil fuel exports from the United States would be reinstituted, and plans to wind down existing production would be laid out. In tandem, a democratized Federal Reserve or similar national Green Investment Bank would ensure that renewable energy got the economic lifeline that it needed during the transition, and current investors in fossil capitalism—particularly public and worker pension funds—would be encouraged to roll over their investments into these green funds. A series of schemes to help workers and communities currently dependent on fossil capitalism transition to equivalent jobs in other sectors would be a key element of the phaseout program. In this manner, a speedy and just transition away from fossil capitalism would take place.

No doubt there would be many specific obstacles that would need to be dealt with in the course of the transition, but the point is to realize that the planned obsolescence of fossil capitalism is not just a necessity according to scientists. It is not just a tangible possibility given public will and mobilization. The liquidation of the fossil capitalist death spiral is also a program that would vastly improve the lives of the vast majority of citizens. Anchored in distributed, renewable energy and supported by renovated models of public power (in both senses of the term), the coming energy transition must generate fresh models of what an energy commons might look like. If the current energy establishment is to be transformed, energy systems—which are fundamental to contemporary global life—must not simply be decarbonized and made more efficient but must equally crucially be used to empower rather than exploit working people, low-income communities, and communities of color. Energy, in other words, must be seen as a commons, not a commodity. As we fight for a future in which the gaping inequalities of the present are overthrown, we reanimate the aspiration for common wealth, a yearned-for ideal that heals the festering economic wounds inflicted on the body politic by unrestrained capitalism in recent decades and, in tandem, revives the fundamental sense of solidarity that gives meaning to human life on this fragile planet.

CHAPTER THREE: A BRIEF HISTORY OF POWER

Speaking before a joint session of the Pennsylvania state legis-lature in the winter of 1926, renowned conservationist Gifford Pinchot outlined a scheme to transform the state's electric grid. The breathtakingly ambitious plan was known as "Giant Power."[171] Drawing on his history of fighting for government planning for the beneficial use of natural resources, Pinchot had commissioned a survey of Pennsylvania's power resources soon after his election as governor two years earlier.[172] Headed by the progressive engineer Morris Llewellyn Cooke, the goal of this survey was nothing less than the production of a blueprint for the rational generation and dissemination of electricity to all the state's inhabitants. Speaking to state legislators, Pinchot suggested that "mechanical energy is the heart of modern civ-ilization . . . We owe the present American standard of living mainly to our use of greater quantities of power per inhabitant than any other people on earth."[173] The impact of this mechan-ical energy had been nothing short of revolutionary, Pinchot argued, since steam had "forced the replacement of individual effort and home industry by industrial organization, for the new steam engine was too big, too expensive, and too complicated to be used except by large numbers of workmen under skilled supervision."[174] Harnessing the state's coal resources had helped power the growth of industry and the massive social transformations that it brought. Yet if steam was the material foundation of modern US civilization, its impact had not been foreseen and planned for adequately. As a result, according to Pinchot, "the discovery of steam was followed by generations of fighting on the part of capital to keep, on the part of labor and agriculture to secure, a share in the rewards of greater production."[175] Now Pennsylvania and the rest of the nation were undergoing a fateful transition from steam power to electricity. The transition from steam to electricity promised to be equally revolutionary in its material and social impact as was the initial introduction of fossil power. "It behooves us," Pinchot told state legislators, "not to let it break upon us unawares, not to permit generations of needless bitter conflict to follow it, but to think out the problems it will create, and to take measures in advance to avoid the long train of struggle and disturbance which followed the last great change in industrial power."[176]

Although we might wish to challenge some of Pinchot's assertions about "mechanical energy," much of what he had to say on that day in 1926 continues to resonate today. Indeed, given the clear necessity of a speedy shift away from fossil fuels in order to avert planetary ecocide, Pinchot's observations about the central role of energy in social life and the need to plan for energy transition seem remarkably prescient. As Pinchot argued, energy generation and distribution play a key—if not completely determining—role in shaping human relations. Every form of energy catalyzes a particular organization of work and division of labor, both within the energy sector and more generally. Shifts in energy generation have consequently been associated with sweeping social, cultural, economic, and political transformations. As humanity moved from gathering and hunting to agriculture, from dependence on wind for power and transport to steam power, and from coal to oil, social relations, structures of feeling, and entire landscapes of being have been radically transformed. Although Pinchot's depiction of a balanced tug-of-war between capital and workers does not make this clear, each of these transformations has tended to concentrate power—in both senses of the term—to a greater degree.[177] If the history of energy revolves around the augmentation of the productive powers of cooperatively organized human labor, over the last two hundred years this massive enhancement of social cooperation has taken place largely at the behest of capital. While workers in some parts of the world have certainly benefited from the social changes wrought by fossil capital, the introduction of new forms of energy also propelled heightened exploitation of labor as well as a global "great divergence" between Euro-American imperial powers and much of the rest of the world.[178] Today, as we struggle to build new, renewably based energy systems, we once again face the question Pinchot outlined to Pennsylvania state legislators nearly a century ago: How might the coming energy transition contribute to new relations of production and exchange that are based on solidarity rather than exploitation? If past energy transitions have tended to intensify the power of elites, can the current energy transition help spark a broader shift toward more egalitarian and democratic social relations?

In order to answer these questions, it is important to consider the factors that have catalyzed shifts in energy regimes in the past. The world has undergone two major transformations in energy regimes since the onset of the Industrial Revolution: the rise of coal in the nineteenth century, replacing a capitalist system fueled by wind and water power, followed by the rise and global diffusion of oil in the twentieth century. Coal, of course, did not disappear in the twentieth century, but it ceased being the predominant source of power in sectors such as manufacturing and transportation. We are, it is to be hoped, in the midst of a third major energy regime shift, although as we saw in the last chapter this transition is not advancing nearly as quickly as it needs to. What combination of factors led to these transitions from one dominant energy regime to another?

The historical record shows that three key interacting dynamics have produced rapid and thoroughgoing energy transition.[179] The first of these is commercial rivalry. Within a capitalist system predicated on incessant competition, those industrialists who can harness the most concentrated forms of energy to power their enterprises will bury their competitors. Myriad examples of this dynamic exist but one clear instance involves increasing commercial investment in coal mining in the nineteenth century, a process that had begun with state-led expansion of coal during the Napoleonic Wars, but that quickly fueled an economic boom in subsequent years as industrialists harnessed the power of coal for manufacturing and transportation.[180] A second catalyzing force in energy transitions, one that is clearly related to the first factor, is geopolitical rivalry, as competition between hostile states propels the adoption of more efficient energy technologies. The classic example of such a competition-stoked transition is the British Royal Navy's switch from coal to oil power in the early twentieth century as it faced an increasing menace from the German navy.[181] The third, and perhaps the least acknowledged, factor catalyzing energy transitions, is popular social mobilization and political conflict around specific energy regimes. Again, there are many dramatic examples of this dynamic but two are particularly worth commenting on. In the first instance, coal miners in western Europe and the United States organized into militant unions in the late nineteenth and early twentieth centuries and disrupted supplies of coal by exerting power over vulnerable choke points in the supply chain like mines and railroad terminals. Their demands included not simply better working conditions, but also a far more sweeping democratization of political, economic, and social life in the oligarchically ruled colonial powers in which they resided.[182] In order to repulse these demands and the militant organizing within the energy sector that gave them bite, elites in government and business switched quickly toward a new petroleum-based energy regime, whose specific material characteristics made mass organizing of the type carried out by coal miners far harder to achieve.[183] Nonetheless, after only a few decades, nationalist uprisings in oil-rich colonized countries in the Middle East and Latin America—movements often led by oil workers—caused serious disruptions to a global energy system overreliant on oil as the predominant source of power.[184] The nationalization of petroleum reserves in countries such as the Soviet Union following the Bolshevik Revolution of 1917, Mexico (1938), Iran (1951), and Iraq (1961) involved confiscation of oil production facilities owned by oil companies based in imperial nations like Britain and the United States. Such acts of appropriation precipitated swift reprisals from Big Oil and the states acting at their behest in the form of economic embargos, clandestine campaigns of regime change, or outright invasion. Nonetheless, this trend toward nationalization continued, with the formation of the Organization of Petroleum Exporting Countries (OPEC) in 1960 signifying a particularly important shift, as a group of formerly colonized nations formed a cartel modeled on the organization of Big Oil in the imperial powers. Efforts to undo these nationalizations have largely failed, and Big Oil has consequently embraced increasingly extreme modes of extraction, whether through expansion into formerly marginal geographical areas or in the form of new, environmentally destructive technologies such as hydrofracking.

During and after its founding, OPEC presented itself as the embodiment of the dreams of emancipation from centuries of colonialism and imperialism of the peoples of the world's oil-producing regions. The formation of OPEC was part of a broader wave of decolonization, one that crested with the Tricontinental Conference of 1966 in Havana (aka the Solidarity Conference of the Peoples of Africa, Asia, and Latin America). This meeting brought together more than five hundred representatives from revolutionary national liberation movements. Its aim was to build the Third World anti-imperialist project begun at the 1955 Bandung Conference and continued by the countries of the Non-Aligned Movement, linking this decolonizing project to a set of explicitly anti-capitalist economic demands.[185] By the early 1970s, these currents generated a series of proposals at the United Nations for a New International Economic Order (NIEO). The demand for a more just global economic arrangement was driven by recognition that conditions had deteriorated for many developing countries during the period after 1945, leading to agitation for the replacement of the Bretton Woods order with a system that would reverse the underdevelopment of the Global South. As the Declaration on the Establishment of a New International Economic Order, a resolution adopted by the UN General Assembly in May 1974, puts it: "The gap between the developed and the developing nations continues to widen in a system which was established at a time when most of the developing countries did not even exist as independent States and which perpetuates inequality."[186] The NIEO was centered on the demand that states should exercise full sovereignty over their natural resources. Such sovereignty included the right to restitution for past exploitation of natural resources during the colonial era, for regulation and supervision of the activities of transnational corporations, and for the right to expropriate foreign property. In addition, the NIEO's foundational document argued that postcolonial nations should be able to set up associations of primary commodity producers like OPEC without fear of economic, political, or military reprisals.

These aspirations were largely represented in the countries of the Global North during the so-called energy crisis of the 1970s not as acts of liberation, with which struggling workers in the United States and the European Union might mobilize in solidarity, but rather as the machinations of duplicitous "Arabs." In fact, the threatened OPEC oil embargo against the United States for its obstruction of a settlement in the Israel-Palestine conflict in 1973 never took effect, but the specter of such an embargo helped legitimate emergency measures by US legislators opposed to a Middle East peace settlement that tightened oil supplies, unleashing a public panic and a surge in oil prices. The "energy crisis" thus provided a pretext and scapegoat for the increase in oil prices long planned by Big Oil.[187] In addition, the ensuing economic crisis helped legitimate the rounds of unemployment, austerity budgets, cuts in social welfare programs, and beefing up of the prison-industrial complex that were integral to the elite project of smashing the Keynesian Welfare State and imposing neoliberalism in the Global North.[188] Today's unending "War on Terror" flows directly out of imperial policies promulgated in the wake of the energy crisis of the 1970s. For example, in the Carter Doctrine of 1980, famously mild-mannered President Jimmy Carter declared the Persian Gulf to be an area of "vital interest" to US "national security" and stated that the United States would use military force to defend those interests.[189] This imperial ideology has unleashed decades of violent mayhem against people in the oil-producing regions of the world while dealing a devastating blow to efforts to build transnational solidarity against exploitation.

Equally tragic, this revivified imperialism has made the hope of collective ownership of the means of energy generation seem nothing more than a fever dream. In the face of these historical defeats, it is essential today to remember the role of mass movements for control of energy resources in reshaping energy industries and in sparking even broader social transformation in the relatively recent past. The challenge we face today is not simply popular control of the energy commons, but rather control dedicated to putting an end to a capitalist organization of energy flows and social life more broadly that is the basis of social injustice and ecological disaster. The capitalist imperative of ceaseless growth on a finite natural resource base is unsustainable, and fossil capitalism must consequently be dismantled. The key to this transition is the insistence that renewable energy resources must be owned and controlled in a manner that ensures that they remain a common good, that they are used in the service of human needs rather than for private profit, and that they are shared equitably around the globe rather than hoarded by powerful imperial states and the elite groups that control them. Access to safe, affordable, clean energy must be seen as a right rather than a privilege. As utopian as this demand for the decommodification and democratization of energy may sound today, the idea of the energy commons has resonated with remarkable frequency over the last century.

GIANT POWER: A PRELUDE

In his 1926 address to the Pennsylvania state legislature, Gifford Pinchot described Giant Power as "a plan to bring cheaper and better electric service to all those who have it now, and to bring good and cheap electric service to those who are still without it."[190] Giant Power, in other words, was a scheme for electrification. Using the possibilities created by technological advances that allowed transmission of electric current over hundreds of miles, Giant Power proposed to link the entire state of Pennsylvania into one huge electric network. The Giant Power Survey Board proposed that large power plants be set up in the western portion of the state, a region rich in coal, with a web of transmission lines being built to bring electricity from these power plants to consumers in both rural and urban regions of the state.[191] Railroads were to be electrified, and other sources of electricity, such as waterpower, were to be integrated into the statewide grid. While this scheme might not sound revolutionary today, Pinchot was clear about the transformative impact that electrification would have for myriad different groups of people in the state. "To the housewife," he said, "Giant Power means the comforts not only of electric lighting, but of electric cooking and other aids to housework as well."[192] It is notable that Pinchot chose to emphasize the gendered impact of electrification, a concern that was shared by other social reformers of the 1920s who sought to rethink domestic spaces in order to liberate women as much as possible from drudgery.[193] But Pinchot's celebration of the emancipatory impact of Giant Power also extended to rural people, urbanites, consumers, and workers. "To the farmer," Pinchot argued, "it means not only the safety and convenience of electric light, but electric power for milking, feed-cutting, wood-sawing, and a thousand other tasks on the farm. To the dwellers in industrial cities it means freedom from the smoke nuisance and the ash nuisance. To the consumer it means better service at cheaper rates. To every worker it means a higher standard of living, more leisure, and better pay." Pinchot and his colleagues in the Giant Power Survey Board were effectively conjuring up a historic bloc of citizens whose diverse outlooks and needs could be unified and mobilized around the intersecting desire for electrification and the transformations in everyday life that it would create.[194]

Giant Power was so revolutionary precisely because it promised to unify an otherwise fragmented electric system. This fragmentation goes back to the origins of electricity in the country, a fact evident from Thomas Edison's Pearl Street Station, the world's first central power plant. Opened near New York's financial district in 1882 and powered by the burning of coal, the Pearl Street Station was used by Edison primarily to generate electricity to run his recently invented incandescent light bulbs, although the plant also distributed heat to nearby customers' buildings. As competitors surged into the new market for electricity, power was used not just to make light but also to run machines formerly driven by steam. But Edison and his competitors used direct current (DC) power, which cannot be transmitted more than a mile and which generates inflexible voltage that is set at the dynamo. Since a light bulb, a street trolley, and a factory machine each require very different voltage to function, a wild profusion of competing private power systems sprang up wherever electricity came into use.[195] In these early days of electricity, there was no notion that there should be a single electrical grid, universally accessible and capable of powering a variety of machines. The boom in the generation of electricity produced a chaotic landscape of competing power companies, with the wealthy opting out by simply setting up their own private generators in the basements of their mansions. City streets were clogged with horse-drawn carts hauling coal while the power lines of competing power companies snaked dangerously overhead.

It was only with the adoption of alternating current (AC) power in the 1890s that this chaos of competing grids was integrated. Unlike DC power, AC power can be transmitted long distances and, even more importantly, its voltage can be transformed using step up/step down transformers.[196] The triumph of alternating current was assured when the Westinghouse Company, armed with Nicola Tesla's AC system, won the bid to illuminate the 1893 Columbian Exposition, producing a dazzling display of power in what came to be known as "White City." Alongside inventions such as the rotary converter, which transformed DC power to AC power and vice versa, this new technological assemblage allowed factories, tram lines, and other enterprises to run their existing equipment with electric power derived from central coal-fired power stations. These new, cooperating technologies sparked a revolutionary transformation of the energy commons: for the first time, power did not have to be produced and consumed individually, where and when it was to be used, but could be generated in one central site, transmitted across many miles and then used in bulk, wherever it was needed and for a virtually infinite variety of tasks, from running railway engines to making lemonade. Together with oil, this new electrical capacity provided the infrastructural framework for a second industrial revolution.

Once these technological roadblocks were overcome, private corporations quickly took monopoly control over the provision of electricity. It was not just that power companies were being built during the age of monopoly capitalism, when robber barons like John Rockefeller and J.P. Morgan almost single-handedly controlled vast swaths of the American economy.[197] In addition, the widely accepted doctrine of natural monopoly also helped support this slide of American capitalism toward oligopoly. In the mid-nineteenth century, the English philosopher John Stuart Mill had argued that the urban gas and water networks being constructed at the time in cities like London were natural monopolies.[198] It would make little economic sense, Mill reasoned, to have multiple, competing systems of pipes bringing water into British cities. To avoid such illogical (and expensive) redundancy, private companies would tend to divvy up cities into exclusive territories. But such arrangements prevented any one company from being able to achieve the economies of scale that would allow them to charge lower rates to their customers. Mill therefore concluded that the establishment of natural monopolies in networked systems providing cities with water and electricity was in the interest of the citizens.

Yet if it was generally accepted that urban power systems should be monopolistic, it did not follow equally that such systems should be in private hands. Adapting Mill's arguments in the context of Progressive-era America, the economist Richard Ely suggested that while natural monopolies might be necessary in urban infrastructure, if they were left in the hands of private corporations they would throttle their customers with exorbitant rates.[199] Ely's critique of private ownership was sparked by a variety of rampant abuses of monopoly power that were raising public ire at the time he wrote. In Seattle, for instance, fire companies were unable to extinguish blazes in some neighborhoods because privately owned water companies had refused to extend the network of fire hydrants to some portions of the city.[200] To combat such abuses of monopoly power, Ely argued, municipal governments should take over the infrastructures that provisioned cities.[201] Ely maintained that public ownership would lead to many benefits for city residents. With infrastructures in public hands, for example, the profits from the provision of gas, water, and other services necessary to urban life would flow to the public in general rather to wealthy shareholders. Democratic oversight of networked systems would ensure far more equitable access for all city residents than would the need to generate dividends for investors. Confronting the idea that the United States was founded on laissez-faire principles and that public ownership of infrastructure was therefore un-American, Ely argued that when the nation was founded, "there were then no monopolies, no collective and non-competitive services where private ownership was dangerous and public ownership a necessary safeguard."[202] If houses had been individually and privately supplied with water and light a hundred years ago, the establishment of natural monopoly corporations at the end of the nineteenth century meant that these necessities of life were now provided collectively and within the public sphere. Therefore, Ely concluded, the public had a right to assert democratic control over this networked commons.

Ely was not alone in arguing for public control of the energy (and water) commons. The critique of private power and the assertion of claims for public control was key to Progressive attacks on the robber barons of the Gilded Age. In his 1895 book *The Coming Revolution*, for instance, the agitator and author Henry Laurens Call argued for public ownership not just of municipal infrastructures like gas and water, but also for the nationalization of railroads, telegraphs, and banks. Corporate control of these "public utilities" must cease, Call thundered, and control of utilities "should pass to the people."[203] Although Call was clearly on the radical end of the political spectrum at the time, his representation of electric light works, gas works, water works and similar infrastructures as public utilities echoed the outlook advanced by more mainstream figures like Richard Ely. Once it became common sense that these infrastructures were public utilities, networks of provision that served the public good, government regulation and even ownership of these infra-structures could be easily legitimated. Three years after the pub-lication of Call's polemic, officials in San Francisco wrote a new charter that committed the city to acquisition of key public utili-ties. A decade later, the Progressive state legislature in Wisconsin passed a bill giving the state's Railroad Commission the author-ity to oversee and intervene in the operations and prices of telephone, telegraph, gas, electric light, water, and power com-panies.[204] Similar laws were passed in New York, where the Municipal Ownership League campaigned for dispossession of the massive business "trusts" that controlled public utilities in the city. These laws, which facilitated public regulation rather than outright ownership of public utilities, were a compromise with powerful business interests. The hope was that the threat of municipal ownership of public utilities would push them to act in the public interest. But after a decade or so, during which power companies grew to be regional rather than exclusively city-based enterprises, the shortcomings of this arrangement became clear. By the 1920s, radical groups such as the Public Ownership League were agitating for democratic control and ownership of essential public utilities.

The Giant Power program built on this history of agitation. From its inception, the campaign was intended to have national ramifications. Morris Cooke, the engineer whom Pinchot tasked with leading the Giant Power survey, believed that the rapid nationwide expansion of electricity in the early twentieth century was following the ominous logic of railroad development in the preceding century. During the nineteenth century, a small handful of tycoons such as Cornelius Vanderbilt had gained control of much of the nation's transportation infrastructure, and then collaborated with other oligarchs such as John D. Rockefeller of Standard Oil Company to establish monopolies in other equally important sectors of the economy.[205] After they eliminated the competition, these monopolies not only charged whatever rates they wanted for the services upon which average people depended, but they also reorganized Wall Street into a national money market in order to generate the vast amounts of capital they needed to function. They then used this capital to buy compliance from politicians throughout the country.[206] These interwoven monopolies were often represented in the popular press of the time as an octopus or a snake. Electricity would follow the pattern set by these earlier monopolies, Cooke worried, unless a state like Pennsylvania could seize leadership and demonstrate how this new resource could be used to benefit the public in general. Other states, he hoped, would follow Pennsylvania's example, until eventually a progressive federal government would take the lead.[207] Giant Power was thus but one element in a bigger campaign that Cooke called "the larger game of building the Great State."[208]

Pinchot shared with Cooke a concern about the trend toward monopoly in the electric sector. For Pinchot, citizens would feel the myriad benefits of Giant Power only if corporate domination of the electric industry could be effectively controlled. In order to dramatize the importance of reigning in corporate power, Pinchot conjured up a dramatic contrast between Giant Power and monopolistic corporate power, which he termed "Superpower." The two, he argued, "are as different as a tame elephant and a wild one. One is the friend and fellow worker of man—the other, at large and uncontrolled, may be a dangerous enemy. The place for the public is on the neck of the elephant, guiding its movements, not on the ground helpless under its knees."[209] Pinchot derived the term "Superpower" from a report published in 1919 by the US Geological Survey, a document that articulated a plan for integrating the country's power-generating capacity into a single large grid to serve the industrialized Northeast.[210] But the Geological Survey report included no guarantees of public regulation or control of this integrated grid, causing alarm among critics already concerned about the increasing monopolistic concentration of the nation's energy infrastructure.[211] "Superpower" was turned in the hands of these critics into a term aptly describing the concentration of immense wealth and power in the hands of a few that the proposed rationalized but unregulated system would help bring about.

In his address to Pennsylvania legislators, Gifford Pinchot described the dangerous implications of the consolidation of electric power in the starkest possible terms: "Nothing like this gigantic monopoly has ever appeared in the history of the world. Nothing has ever been imagined before that even remotely approaches it in the thoroughgoing, intimate, unceasing control it may exercise over the daily life of every human being within the web of its wires."[212] In seeking to convey the potentially destructive impact of Superpower adequately, Pinchot reached repeatedly for metaphor, first describing it as a rampaging elephant but then turning to an even more sinister, vampiric image: "[It] is as though an enchanted evil spider were hastening to spread his web over the whole of the United States and to control and live upon the life of our people."[213] For Pinchot and colleagues of his like Morris Cooke, the rapid consolidation of Superpower meant that there was little time to create an alternative integrated electrical grid that would function not for private profit but in the public interest.

Given Pinchot's scathing denunciation of Superpower, it is perhaps surprising that he did not explicitly advocate public ownership of its antithesis, Giant Power. The Giant Power plan, he said, "takes no account of public ownership. It proposes to deal with facts as it finds them, and does not even raise the question."[214] Public regulation of electric monopolies is essential, but that regulation, Pinchot argued, may be carried out at either the state or the federal level. Pinchot's position on public regulation is a product of a surprising moment a quarter of a century earlier. Just before the end of the nineteenth century, a movement for public power on the municipal level had caught fire: from 1897 to 1907, 60 to 120 public power systems were formed each year as a result of municipal referendums, a rate of growth twice that of private power.[215] The owners of private power corporations, terrified by this growth of public power, searched feverishly for ideas that would beat back this swelling tide of municipalization. In 1898, Samuel Insull, the former private secretary of Thomas Edison who had moved to Chicago and established an electricity empire in the Midwest, used his presidential address to the electric industry's annual convention to articulate a proposal to defang the movement for public power. "It seems to me," Insull told his listeners, "that the claim that municipal operation is the universal cure for all diseases for which electric lighting companies are supposed to be responsible, merely proposes the substitution of the political in the place of industrial management. This raises the question: is the administration of municipal affairs in the various cities of this country so economic as compared with the management of private industries, and the class of service rendered so efficient, as to justify the increase of the burdens already imposed upon municipal government?"[216] This was clearly a disingenuous question since Insull and his fellow corporate titans did not believe that public administration of electricity (or anything else, for that matter) was more efficient or economical than corporate power. Indeed, Insull had already stated in his speech that "[we] all realize, from the close attention we have given to our own affairs, that self-interest and the necessity of getting a return on our investment are the first essentials to the economic administration of large enterprises."[217] Insull's arguments are strikingly similar to those that would be advanced during the current neoliberal era for the superiority of private over public provision of essential services, a doctrine that has led to repeated waves of privatization since the 1970s. This similarity makes the conclusion that Insull goes on to draw all the more remarkable: "It appears to me that a correct division of power and responsibility requires political government to control private industrial management. Where political government and industrial management are merged into one interest, the power of control is seriously impaired, since a political administration cannot be reformed without overturning the party in power."[218] Rather than calling for the dissolution of public power, Insull called for direct public regulation of the private corporation, rather astutely noting that the only means of reforming a public entity charged with controlling provision of municipal electricity, water, and other supplies in a representative democracy is through infrequent (and potentially corrupt) elections. Insull went on to argue for the granting of exclusive franchises to private companies operating in specific publicly regulated terrain, reasoning that the resulting security of operations would allow the companies to raise capital at reduced interest rates, lowering the total cost of operation, which would in turn lower the price of service to users.[219] Insull's advocacy of stringent public regulation of the electric industry, and his arguments against competition and the free market, shocked and scandalized his peers. Indeed, he could get no one to respond to his address. But, not many years after his iconoclastic speech, the National Civic Federation—one of the most important organizations of the day, with board members such as Andrew Carnegie and acolytes of J.P. Morgan—established a Commission on Public Ownership that called for a legalized system of electric monopolies under public regulation.[220] In 1907, Wisconsin and New York followed this report's recommendations, becoming the first states in the nation to establish utility commissions charged with supervising electric companies. In doing so, they took regulatory authority out of local hands and blunted the movement for municipally based public power. As Milwaukee mayor Daniel Hoan put it in 1907, "[It's] supposed to be legislation for the people. In fact, it's legislation for the power oligarchy."[221]

Publicly regulated private power companies—which in the last chapter I called investor-owned utilities (IOUs)—continue to provide the great majority of power consumed in the United States today. In other words, Insull's crafty strategy to head off the movement for public power is still alive. Yet already by the time of Pinchot and Cooke's Giant Power plan, it was clear that public regulation of private power was not working adequately. It was not just that state commissioners, appointed by governors in most cases, tended to be firmly in the pockets of the big power companies, meaning that rates for consumers were negotiated between commissioners and power company representatives behind closed doors.[222] In addition, there was an obvious and jarring problem of scale. As Pinchot put it in his address on Giant Power, "It is axiomatic that to be successful and effective the regulating machinery must cover the same ground as the thing it regulates. Regulation of a nationwide electric combination by a state alone consequently carries with it such inherent difficulties and such disadvantages, from the public point of view, that nothing less than the wholehearted co-operation of the companies and the states can give it even a reasonable prospect of success."[223] But cooperation on the part of the power companies was rarely forthcoming. This fact led Pinchot to suggest implicitly that nationalizing power might be an inevitable necessity: "I venture to say that if the people of the United States ever turn to the nation-wide public ownership of electric utilities, it will be because the companies have driven them to it. It will be directly and only because the utility companies have so opposed and prevented reasonable and effective regulation by the states and by the nation that the only choice left was between servitude to a gigantic and unendurable monopoly and the ownership and operation of that monopoly by the people."[224] Pinchot's warning about the necessity of public power given the intransigent behavior of corporate power companies was to resonate increasingly loudly in subsequent years.

The Giant Power plan was explicitly concerned with the rationalization of energy infrastructure, but it was always about far more than that. The fight between Giant Power and Superpower was over who would control the terms of a transformation that was at once gigantic in scale while also affecting the most intimate elements of everyday life, for electrification was both a massive infrastructural project and a means of revolutionizing people's quotidian routines, emotions, and aspirations. Pinchot sought to articulate the deep cultural and social impact of electrification in terms that were both analytic and speculative. He argued that the United States was on the edge of a shift that would not only be manifold but that might redeem some of the destructive transformations brought about by fossil capitalism during the last century. "Steam," Pinchot argued, "brought about the centralization of industry, a decline in country life, the decay of many small communities, and the weakening of family ties."[225] Like many other Progressives, Pinchot was keenly aware of the destructive impact of monopoly capitalism on rural life in the United States. His Giant Power plan, he accordingly hoped, would help reverse the dismantling of rural life: "Giant Power may bring about the decentralization of industry, the restoration of country life, and the upbuilding of the small communities and of the family." Giant Power in fact found some of its most ardent supporters among the farmers of Pennsylvania. Writing in the *Pennsylvania Grange News*, the main organ of public opinion for the state's farmers, Progressive economist Basil Manly denounced the "alarming decline of American agriculture" and argued that control of the nation's resources by monopoly capital "must be destroyed and a new system of cooperative distribution, for service and not for profit, must be [built] in its place."[226] The editor of the same newspaper, John A. McSparran, who was also Master of the state Grange, attacked "the outrageous profiteering of the big combinations of capital that defy law and public sentiment," called for large concentrations of wealth to be taxed "out of existence," and, speaking of the private electric utilities, said "we have tried regulation, and these giant corporations laugh in our faces."[227] Giant Power, it was hoped by these and other social critics and political activists, would smash capitalist monopoly power. In addition, it would also upend the character and geography of American civilization. "In a steam-driven civilization the worker must go to the power," Pinchot argued, "but in an electrically driven civilization the power will be delivered to the worker."[228] To Pinchot, the implications were nothing short of revolutionary: "Steam makes slums," he declared, while "[electricity] can replace them with garden cities."

There was, of course, more than a hint of technological determinism in Pinchot's arguments for the civilizational transformation to be wrought by Giant Power.[229] Energy transitions, such as the one Giant Power was to catalyze, are always a result of interacting material, political, and economic forces. But the limits of Pinchot's arguments were never really tested since entrenched interests shot down the Giant Power plan. Although the scheme got widespread and often positive coverage in the press, the electric industry unanimously denounced the proposal, arguing that the limitations on expansion of the grid were commercial rather than technical: with increasing access to capital, industry spokesmen argued, power lines would expand to incorporate rural areas in Pennsylvania and beyond.[230] Other business groups joined in the attack, and the Pennsylvania state legislature ignored or smothered Pinchot's proposals. Even the farm organizations that had once roundly denounced monopoly capital pivoted to support measures that placed the burden of rural electrification on customers rather than on private utilities. Nevertheless, while Pinchot's plan for Giant Power was defeated in Pennsylvania in the 1920s, the ideas and political energies mustered under its banner were to provide a direct inspiration for many of the initiatives that flowered a decade later during the New Deal. Giant Power should therefore be seen less as a failure than as a prelude to public power. The ideas and arguments advanced by Pinchot and his comrades remain important today as we struggle for a just transition to renewable power. Excavating the intense social struggles of the last century over the energy commons, we reanimate our sense that alternative dispensations and alignments of power were feasible in the past and are possible in the present. Looking through the archives of electrification, we recall the fact that public power has been not just an idea but also a reality for which masses of people fought, and we reenergize struggles in the present for people's power.

SPEAKING TRUTH TO POWER

In 1932, Carl Thompson, the cofounder of the Public Ownership League, published a bombshell of an exposé called *Confessions of a Power Trust*.[231] Thompson's muckraking text summarized the hearings of the Federal Trade Commission (FTC) into the financial structure, methods, and operations of the nation's electric power utilities, an investigation carried out in the mid- to late 1920s. Much of the book consists of extended quotations from the testimony of power industry insiders before the FTC. *Confessions of a Power Trust* dwelt in great detail on the industry's many shockingly corrupt dealings, but it was the revelations concerning the power utilities' propaganda campaigns that truly scandalized public opinion. "To carry out their plans," Thompson wrote, "required changes in public thought and action, and, therefore, a molding and recasting of public opinion that were almost revolutionary."[232] In order to mobilize public consent for private power, the industry carried out a systematic campaign of advertising in local newspapers in order to sway their coverage of issues relating to power. The details of this campaign are laid out forensically by Thompson, who discusses items such as "Exhibit No. 637" from the FTC hearings, which "outlines 'The Proposed Public Utilities Advertising Association's'" campaign. When newspapers did not comply with these campaigns, Thompson documents, the power trust purchased them outright, leading to a concentration of ownership and control in newspapers similar to the trend toward monopoly evident in other areas of the US economy.[233] The Power Trust also bought off college professors and publishers so that textbooks reflected the interests of the power trust alone; as the oil industry was to do subsequently, it paid for distribution of curricular materials to schools that naturalized private power; and it essentially bribed civic organizations like Boy Scout troops, church groups, and women's clubs to express support for private power.

As if this campaign to coopt the organs of public opinion in order to manufacture consent for private power were not scandalous enough, the utilities did not pay a dime for it. Thompson quotes the testimony of J. B. Sheridan, manager of the Missouri Committee on Public Utility Information, who told the FTC that, "In this case there is really no expenditure by the companies; they merely allot a certain sum out of their annual gross earnings to educate the people in the processes of their operation and charge that sum to operating expenses."[234] Since utility company profits were fixed by law to 6 to 9 percent of their capital investments and operating costs, the more expenses a utility claimed for advertising, the more money it stood to make: costs were simply passed on to consumers through rate increases. In this way, a sweeping culture war to "change public consciousness" was ironically paid for by the public. In his testimony, J. B. Sheridan stated that the Public Service Commission was quite happy to approve these efforts to "educate" the public about the industry.

How to get the word out to the public about this gargantuan culture war, and about the rapacious and corrupt dealings of the Power Trust that the utilities' propaganda campaign helped obscure? Perhaps the most interesting effort to speak truth to the Power Trust grew out of the Federal Theater Project (FTP), an organ of the Works Progress Administration established in 1935 under the auspices of the New Deal. Under the direction of the pioneering female theater director and writer Hallie Flanagan, the FTP adapted experimental dramatic techniques developed in Russia during the Bolshevik Revolution and then reworked by such radical European playwrights and dramatists as Bertolt Brecht and Erwin Piscator. These techniques melded radical political messages with formal strategies that disrupted the naturalism that had long prevailed on stage and in literary culture in general. After her appointment as national director of the FTP, Flanagan helped form the New York Living Newspaper Unit, which employed out-of-work journalists and theater professionals to produce plays inspired by current events. Although Living Newspaper productions employed a variety of techniques, typically they involved a team of journalists researching a topic to gather factual information for a play, information that was then turned over to a group of editors and writers who would assemble these documents into a narrative. Living Newspaper productions used experimental dramaturgy such as quick scene and set changes, projection of settings and statistics, the use of a loudspeaker to narrate events, and abrupt blackouts and harsh spotlights.[235] But this was not art for art's sake: the aim of Living Newspaper productions was to render controversial and complicated social and political issues relevant and intelligible to the general public. The group's initial productions were highly controversial among the nation's political elites. The unit's first play, *Ethiopia*, which documented and criticized the Italian fascist invasion of Ethiopia, was effectively banned when the federal government issued an edict prohibiting the impersonation of heads of state on stage shortly before the production was to debut in 1936. Subsequent plays produced by the FTP took aim at government agricultural policies and satirized big business, generating even more ire in Congress. In 1937, the Living Newspaper play *Power* opened in New York. The play, which was billed as "a thrilling dramatization of modern industry," presented to live audiences many of the details of rapacious corruption on the part of the electric utilities itemized in *Confessions of a Power Trust.*

Power begins with a projection on the front curtain of the words "The Living Newspaper presents POWER."[236] The word "power" grows larger while the other words fade out. As if the double meaning of the word was not made clear enough by this projection, a loudspeaker-narrator then describes a power outage that took place in New Jersey in 1936. The stage is thrust into darkness, and a series of different people appear in quick succession, their lives thrown into crisis by the loss of electricity: a doctor and nurses who must carry out an operation using only flashlights for illumination, an Irish mother who worries that her child's sickness will worsen after the heat goes off, a panicky airport air traffic controller, and a driver who disconsolately protests that he didn't see a pedestrian he has apparently run over because of the darkness. After establishing the importance of power to everyday life in the United States, the play then looks back at the history of electricity. The narrator introduces a series of scientists; it is a list that culminates with Thomas Edison, who exclaims as he illuminates the stage with a light bulb: "The happiness of man! I know of no greater service to render during the short time we live!"[237] A group of businessmen then rush in, surround Edison, and demand that he sell them the rights to market this miraculous new technology. The stage directions call for "almost incoherent" excited speech and ad-libbing by this scrum of businessmen, out of which the words "money," "profits," and "thousands, millions, billions" can be heard as they try to wrestle the light bulb out of Edison's hands. While this scene may appear to hammer playgoers over the head with its contrast of the pursuit of science for the public good and the zeal for private gain, like other scenes in *Power*, it is based on a true story: Thomas Edison, who not only invented the incandescent light bulb, but was also the first to conceive of selling electricity itself, was pushed out of business by the Wall Street financier J.P. Morgan, who took over Edison's General Electric company in 1892.[238] The subsequent consolidation of the power business, with GE controlling three-quarters of the electric market in the country, is depicted in *Power* through the appearance of a financier who admonishes the quarrelling businessmen, telling them that competition is ruinous for the bottom line, and offering to buy out their holdings in order to form one big corporation. Thus *Power* narrates the formation of monopolistic Superpower during the late nineteenth and early twentieth centuries with admirable clarity and economy.

The upshot of this trend toward monopoly is laid out through the play's depiction of the infuriating experiences of a hapless character known only as CONSUMER. This modern everyman is depicted as he grapples with the obscurity of electric company rates. Compared to gallons or yards, the play asks, what exactly is a kilowatt-hour, the unit in which payments for power are measured? The consumer has no idea, but even when he does decide that the rates are too high, his efforts to find lower rates elsewhere are stymied by the consolidated character of power: as a company manager tells him, "there is no other fellow" to bargain with for lower rates. At the same time, a projection above the stage announces "*M is for Monopoly.*"[239] When consumers try to unite and fight exorbitant rates in court, they lose as a result of the Power Trust's legal muscle and are charged fees for the court case. A series of scenes follow that unpack the corrupt structure of "holding companies" through which corporate bosses like Samuel Insull accumulated dizzying numbers of nominally independent power companies under their control. *Power* employs a powerful conceit to demonstrate the Ponzi scheme character of these holding companies, using stacked boxes on stage to represent the electric operating companies and the financial holding companies within which they were subsumed as a great pyramid structure. The unmasking of corruption culminates in a brilliant and hilarious scene in which an actor representing Insull demonstrates to a gawking consumer-investor "how to make a lot of money." Insull buys various different companies using money he has borrowed from other companies that he owns, transferring this money from one pocket to the other while the worth of his investment grows astronomically. To the consumer-investor's indignation, Insull announces that he holds a controlling percentage of shares and can determine exactly how much he will pay himself, despite the fact that he has invested "not one red cent" in the enterprise.[240]

Power concludes its scathing attack on the blatant corruption of the Power Trust by documenting the industry's debasement of public opinion through the propaganda campaign detailed in *Confessions of a Power Trust*. The play's most lacerating satire concerns the corruption of academia by the Power Trust. As a file of professors enters wearing cap and gown, the loudspeaker-narrator intones, "In the colleges . . . the following members of a notoriously underpaid profession find extra-curricular employment as lecturers, editors, and advisors for public utility corporations."[241] As the professors walk across the stage, the loud-speaker announces, "a professor from Rensselaer Polytechnic Institute—eight thousand dollars; a professor from Yale—eight thousand one hundred and seven dollars and twenty[-]five cents [. . .]."[242] When the loudspeaker confronts these eminent schol-ars, saying "Gentlemen—what do you think of the doctrine of municipal ownership?" the stage directions dictate, "In unison, mournfully, they all shake their heads, indicating that they con-sider it entirely hopeless." Not to be deterred, the loudspeaker continues, "What do you think of Government supervision and control of the abuses we have shown to exist in the public util-ities?" "Again," the stage directions indicate, "they shake their heads." When, finally, the loudspeaker asks, "What do you think of a nice juicy steak smothered in onions?" the following reaction is elicited: "The PROFESSORS nod their heads energetically, with broad grins on their faces and very much interested."[243] The gag is crude, but so were the politics and the material interests that drove them.

After documenting and denouncing the corrupt dealings of the Power Trust, *Power* goes on to scrutinize the question of public power. A dialogue between a young girl and her father that occurs nearly halfway through the play sets the tenor of this exploration. The girl asks her father where electricity comes from. Very quickly her questions demonstrate that the father, like most of his contemporaries, knows little about how power is "generated." When the girl asks why the government doesn't produce their electricity since it is so important to modern life, the increasingly exasperated father replies that this "would be competing with private business, and, besides, everybody knows that the Government wouldn't be efficient."[244] The girl then asks who runs the post office; when her father answers that the government does because it's too important a service to let anyone else provide it, the girl asks, "[Don't] you think electricity is important?" Through this scene of "childish questions," the carefully cultivated commonsense belief that it makes sense to entrust a key public service to corporate interests that are bent on private gain is dismantled. The final turn of the screw comes when the girl asks who the government is, to be told that it's "you and me—the people." After she asks, "[Does] the company own what the people need?" and is told that it does, the girl replies "Gee, Daddy, the people are awfully dumb." But people are not dumb: portions of the US populace were convinced through a pro-corporate propaganda blitz that the Power Trust was a natural arrangement. *Power* demonstrates how little sense this situation makes, and how it leads to the gouging of the average consumer. While the dialogue between the girl and her father may not appear to be very subtle, it does obliquely and slyly raise the question of expropriation and socialization of the means of production. Surely, following the girl's logic, the company should not own what the people need—the people should own what they need. Again, this analysis is true to the moment: it was mass working-class struggle that threatened to delegitimize not just certain industries during the Great Depression, but also private property and capitalism itself.[245] Without this radical contestation of the very terms around which society was organized, there would have been no New Deal reforms.

The bulk of the second half of the play addresses the question of public power by focusing on the Tennessee Valley Authority (TVA), the legality of which was under scrutiny by the Supreme Court at the time *Power* was produced. The play's radical orientation toward a contentious present and a future still very much in play is apparent throughout this second half; while not involving audience participation to the extent of Bertolt Brecht's radical *Lehrstücke*, or "learning plays," *Power* nonetheless challenges the idea of a neutral representation, offering contending interpretations of public versus private power and asking theatergoers to draw their own conclusions. This strategy of presenting openly conflicting views is clearest in the play's finale, in which we hear businessmen complaining that the TVA is "un-American" and judges arguing that it's unconstitutional, assertions that are punctuated by the demands of farmers for light and workers for power. Yet before these exchanges take place, *Power* has provided theatergoers with a sense of the visceral impact of living without electricity on rural people. In one scene, for instance, the wife of a farmer who struggles to read by the dim light of a kerosene lamp tells him that he is going blind because they lack power. The woman challenges her husband's passivity using sarcastic humor: "[Suppose] they told you you couldn't have any air," she asks him, "would you stop breathin'?"[246]

Power relates the story of the establishment of the TVA through snapshots of popular struggle, shuttling back and forth between farmers who are stuck without electricity because they cannot afford to pay for the poles and wires necessary to bring current out to their isolated homesteads, and city dwellers groaning under exorbitant utility company rates. The emphasis gradually shifts from feelings of powerlessness to self-empowerment. For example, residents of a small town meet to discuss how their lives would be transformed by electrification; when the high cost of infrastructure is brought up, a "chairman" informs them that they can borrow money from the federal government at low interest in order to build their own power plant.[247] When the private power companies try to run wire first in order to muscle out the government, a farmer cross-examines the linesmen who are stringing wire and then chases them away when he understands the ruse. Farmers even take to sabotage by chopping down and burning poles put up by private power companies. These struggles over the TVA eventually end up at the Supreme Court, where the majority opinion is delivered by Chief Justice Hughes: "Water power, the right to convert it into electric energy, and the electric energy thus produced constitute property of the United States."[248] A victory parade on stage is cut short when the power companies appeal this decision and an injunction is issued banning the TVA from constructing new power lines and servicing new customers. *Power* concludes with the issue of the fundamental constitutionality of the TVA still in question. The final words of the play note, "[The] foregoing finale is subject to change when the TVA issue is finally decided by the Supreme Court." *Power* thus apparently leaves the fate of public power in the hands of an elite judicial decision. Yet through its extensive documentation of the popular struggle around the TVA, and through its efforts to galvanize greater struggle in the present, *Power* reminds us that it is mass mobilization rather than a judicial ruling that is the foundation for public control of power. Indeed, one of the lessons of the 1930s was that a well-organized mass movement could overwhelm obstructionist conservative opposition from the courts.[249] The apparently open-ended conclusion to *Power* should thus be read not as an expression of resignation but as an injunction to popular action.

ABUNDANT LIFE: THE NEW DEAL AND PUBLIC POWER

Late in the summer of 1934, President Franklin Delano Roosevelt stood in front of the works at the Grand Coulee Dam on the Columbia River in Washington State to deliver an address about the social significance of the power that the dam was to generate upon completion. For Roosevelt, Grand Coulee's awesome power was a vital element in achieving the vision of "abundant life" that undergirded the New Deal in general. In his speech at the dam, Roosevelt painted a picture of electricity distributed so cheaply that all Americans would be able to afford it: "We are going to see, I believe, with our own eyes electricity and power made so cheap that they will become a standard article of use, not only for agriculture and manufacturing but also for every home within reach of an electric light line. The experience of those sections of the world that have cheap power proves very conclusively that the cheaper the power the more of it is used."[250]

Electricity is almost universally available today in the United States. Indeed, it is nearly taken for granted, but when FDR spoke at Grand Coulee that was far from the case. Over twenty-six million people, roughly 20 percent of the population, lacked electricity in 1934.[251] Most of these people lived in the nation's rural areas, where farmers and others still lit their homes with oil lamps and were unable to use any of the laborsaving appliances that were becoming increasingly widespread in urban homes of the day. FDR's goal in ordering the construction of Grand Coulee and similar dams was ultimately to electrify the entire nation, but in doing so he wished not simply to extend the electrical grid to rural people. For contemporary commentators, FDR's electrification program was prompted at bottom by an effort to salvage the US's political and economic system.

Writing of FDR's Grand Coulee speech and of the challenge of national electrification, influential contemporary journalist and science writer Waldemar Kaempffert argued, "A Fascist or Communist society would have little trouble dealing with the problem thus presented. The government would simply draw up plans and issue an order. But our government happens to be a democracy—one of the few surviving democracies in the world. It has a tradition of individual freedom, and with it a laissez-faire policy that makes it possible for energetic opportunists to become millionaires but also gives the economically strong an enormous advantage over the weak. Can it plan its own future? Can it cope with the social and economic issues raised by energy generated in bulk and used collectively?"[252] For Kaempffert (and, he suggests, for FDR), electrification was a critical challenge for the United States, which lacked the centralized authority that might allow a fascist or communist state such as Nazi Germany or the Soviet Union to launch nation-spanning infrastructural projects. But in addition to shining a glaring light on the United States' fragmented, inertia-filled liberal democratic political system, the challenge of electrification also underlined the inequities and dysfunctions of its capitalist economy. While it may sound counterintuitive today, after a carefully orchestrated propaganda campaign lasting decades has made the idea of the efficiency of the free market seem quite natural to most Americans, for Kaempffert it made perfect sense to segue seamlessly from the tendency of free markets to concentrate wealth in the hands of a few to the question of the US's capacity to provide power for all of its citizens. What were the material conditions that led to these questions about the energy commons and the viability of capitalism and liberal democracy?

These were the concerns that animated FDR's speech at Grand Coulee. In his discussion of that speech, Waldemar Kaempffert points to the fact that twenty-seven million people were living on farms, very few of whom had access to electricity. This was because, as Kaempffert argued, power companies had no incentive to connect rural areas to the grid. "Power companies," Kaempffert wrote, "are not philanthropic institutions. Investors look to them for dividends. If there is no market in a sparsely populated territory it is ruinous to electrify it, and this because it costs ten times as much to distribute as to generate energy."[253] The logic of the market, in other words, not only militated against the public interest, but also constitutively excluded a significant segment of the public from the infrastructure of the nation. In the face of this contradiction of capitalism,

Kaempffert argued, "the profit motive must be subordinated to the obvious social duty of spreading the use of electricity." As we have already seen, for Kaempffert this was not just a question of social justice but of the fundamental viability of American culture, whose chaotic free market system appeared increasingly like a relic of an age before the rise of the large infrastructural networks of the Second Industrial Revolution, networks that seemed to call for organization and planning that centralized political systems such as fascism and communism were proving adept at creating. Government-owned hydropower complexes like the Grand Coulee Dam, Kaempffert suggested, would allow the US government to cope with these challenges by permitting it to determine wholesale costs of electricity provision, and then mandating that investor-owned public utilities distribute power to customers at such rates. Moreover, Kaempffert stated, "[If] the company declines to meet the rate [...] the President maintains that a community has a right to build and operate its own plant."

FDR's administration backed up these promises during the New Deal era by providing low-interest loans through the Department of Agriculture's Rural Electrification Administration (REA), created in 1935. Rather than actually building the infrastructure that brought electrification and economic development to rural areas itself, the federal government empowered state and local government, nonprofit organizations, and farmer cooperatives through long-term, self-liquidating loans.[254] As the play *Power* documents, the rural power cooperatives supported by the REA built on experiments developed in the South under the auspices of the Tennessee

Valley Authority. Small groups of rural townspeople or farmers created cooperatives, with each member contributing an initial membership fee to cover the construction costs of the power lines and then paying installments in a monthly bill. Citizens were thus empowered to own and manage their own system rather than being dependent on the profit-driven calculations of private power companies. The radical democratic rhetoric of the REA, with its emphasis on member ownership and control of power, stood in stark contrast with the exclusionary and oligarchic control over electricity wielded by the big trusts of the day, but also with centrally controlled political systems such as communism and fascism. The REA had a transformative impact on the energy commons in the United States. In 1934, when FDR delivered his speech at Grand Coulee, only 10 percent of rural residents had access to electricity. By 1939, the REA had made loans totaling $3.6 million and had worked with 417 cooperatives serving 270,000 households.[255] Today, electricity is universally available in the United States, and 75 percent of the US landmass gets its power through cooperative utilities set up under the auspices of the Rural Electrification Association. These co-ops today have over forty-two million members and $164 billion in assets.[256]

In the midst of today's fight for decentralized forms of Energy Democracy during the transition to renewable energy,[257] it is worth remembering not simply the long struggle for rural electrification but the specific *form* through which the countryside was wired: member-owned and controlled co-ops. In their day, these co-ops were a striking example of the energy commons inasmuch as they were a product of direct democracy, existing alongside the state rather than wholly subsumed by top-down, technocratic governance. Rural co-ops, it is true, would not have existed without the forms of economic and political empowerment extended to them by the federal government, support which made their growth in the face of obstruction by powerful private power companies possible. Yet, nonetheless, they were relatively autonomous from the federal government and were correspondingly responsive to the felt needs of rural people. The growth of rural electric co-ops is all the more notable given the fact that the state had long been represented on the Left in monolithic terms, as a force standing in opposition to the popular masses. For the Leninist tradition, class contradictions were located *between* the masses and the state: the struggle of the masses was therefore seen as the creation of a dual or alternate power that would eventually smash and replace the bourgeois state.[258] Surprisingly, as the radical theorist Nicos Poulantzas argued, there is a deep affinity between this Leninist tradition of state theory and social democracy, which aimed to replace bourgeois leaders with enlightened radicals, who would then bring socialism to the masses from above.[259] Both of these traditions were distrustful of direct democracy and popular initiative. Both led to forms of authoritarianism. For Poulantzas, writing in the wake of the uprisings of 1968 and the struggles against authoritarianism that unfolded in the following decade, the key question was: "How is it possible radically to transform the state in such a manner that the extension and deepening of political freedoms and the institutions of representative democracy (which were also a conquest of the popular masses) are combined with unfurling of forms of direct democracy and mushrooming of self-management bodies?"[260] This question continues to resonate today. Indeed, it is perhaps the key question the Left confronts as the climate crisis pushes us to go beyond both the disempowering legacy of Stalinist statism, on the one hand, and the disengagement of some autonomist traditions, which believe in changing the world without taking (state) power, on the other.[261] In the face of these antithetical but equally disempowering options, it is important to remember examples like the rural electric co-ops of the 1930s, which offer an example of precisely the combination of a deepening of representative democracy combined with new organs of popular self-management. Today, the majority of these co-ops are, alas, run in the same opaque, top-down manner as the for-profit public utilities, and these co-ops remain dependent on aging coal-fired power plants as well as the big dams built during the New Deal era. But the prospect of decentralized, democratically controlled renewable energy generation has begun to galvanize many co-op members into reclaiming public power—and kicking the utilities' fossil fuel habit.[262]

CHAPTER FOUR: THE ENERGY COMMONS

Fossil capital has been granted immense power, producing life-giving heat and light but also plunging communities into darkness when they fail to return outsize profits. In 2011, DTE Energy, the investor-owned utility (IOU) that controls southeast Michigan's energy infrastructure, repossessed one thousand streetlights from Highland Park, a city in the larger metropolis of Detroit.[263] The city was left in the dark. Like many other Black-majority cities across Michigan, Highland Park was struggling at the time with capital flight and spiraling levels of austerity. Once home to Ford and Chrysler auto assembly plants and the well-paying jobs that they generated, Highland Park had seen its fortunes crash in the 1990s and the 2000s as the automakers shipped jobs abroad. Now, half of the residents of Highland Park had trouble paying their monthly electric bills, and a quarter had experienced a shutoff of gas or electricity—often during Michigan's cold winter months.[264] When DTE took the lights, Highland Park owed $4 million in electricity bills, a situation likely to be aggravated by the rate hikes the utility wanted to impose to support its existing coal-fired power plants, to build new fossil fuel plants, and to pay the utility's chief executive his $5.4-million annual salary. The repossession of Highland Park's streetlights was part of a broader crisis of public assets: across Michigan, communities struggled as control of key public infrastructure like the water department and the school system was stripped from them by undemocratic emergency-management czars.

The taking of the light in Highland Park is part of a new, global round of enclosures in which common assets are stripped from the public.[265] For radical critics of capitalism such as the historians Silvia Federici and Peter Linebaugh and the geographer David Harvey, enclosures are one of the dominant forms of contemporary capital accumulation.[266] According to these activist scholars, critics of capitalism have mistakenly followed Marx's analysis of what he famously termed "primitive accumulation," which sees enclosure as a kind of original violence that kick-started the capitalist system.[267] Enclosure, Marx argued, was foundational to capitalism since it allowed powerful landlords to accumulate wealth by dispossessing the peasantry of the land they farmed collectively, replacing such feudal social relations with more lucrative forms of enterprise such as the production of wool. As the sixteenth-century English philosopher and statesman Sir Thomas Moore put it, "sheep, which are naturally mild, and easily kept in order, may be said now to devour men and unpeople, not only villages, but towns."[268] The accumulated capital produced by enclosure of common lands was used to support expansion of industrial production domestically and of the transatlantic slave trade and colonialism abroad. Enclosure thus refers to a global process of violent extraction. But the key thing is that enclosure did not cease once common lands in Britain had all been opened up and capitalism had been established as an economic and political system. Contrary to what Marx argued, the predatory stripping of common assets around the world never stopped.[269] In fact, it has intensified. The neoliberal era that began in the 1980s has seen a massive expansion of attacks on the commons, both in the form of the shifting of formerly public assets such as school systems into the private sphere in rich countries, and through extensive land grabs in areas hitherto relatively autonomous from the capitalist world system such as parts of sub-Saharan Africa.

Resistance to the new enclosures has become a central feature of social struggles over the last few decades. For instance, in 2000, the people of the city of Cochabamba in Bolivia rose up after the World Bank insisted that the government hand over control of municipal water supplies to *Aguas del Tunari*, a conglomerate controlled by the US-based multinational Bechtel Corporation. The new owner of Cochabamba's water demanded steep and sudden rate increases of double or more for poor consumers in order to finance the double-digit profits demanded by the companies.[270] The conglomerate even proposed to tax water that people caught in barrels as the rain flowed off their roofs.[271] The people of Cochabamba rose up in protest, occupying the center of the city and forming a grassroots participatory organization called the *Coordinator for the Defense of Water and Life* that shut the city down and demanded a rollback of the water privatization measures.[272] Under pressure from the water conglomerate and international authorities, the Bolivian government declared martial law and tried to suppress the protests with riot troops, measures leading to mass arrests, hundreds of injuries, and the death of a teenage boy as conflicts erupted on the barricades the citizens had set up around the city. Protesters held fast in the face of state repression, however, and on April 10, 2000, the Bolivian government reached an agreement with the *Coordinadora* that ultimately not only reversed the privatization of the city's water but also catapulted Evo Morales and his Movement for Socialism (MAS) into power in the country.

This victory for popular mobilization in Bolivia was a key moment in resistance to the new round of capitalist enclosures carried out during the age of neoliberal hyper-capitalism. The defense of the commons through new forms of participatory organizing resonated around the globe in the following years. In 2013, for instance, Turkish activists protesting government plans to pave over Istanbul's Taksim Gezi Park described the park itself and various other urban spaces that the government's neoliberal policies tried to confiscate for private profit as a "commons." The Turkish activists called the form of self-government developed during their occupation of Gezi a "commune," one that involved not just a sit-in but also food distribution, a medical center, and an autonomous media collective.[273] Grounded in a determination to defend common space from enclosure, the Gezi protest shared a commoning ethos not just with the Cochabamba Water Wars but also with similar movements around the world, from the resistance of the Zapatistas to the neoliberal tenets of the North American Free Trade Agreement (NAFTA) in Mexico beginning in 1994, to the Occupy movement that began in New York and spread across the United States, and to the Indignados movement in Madrid and Spain in 2011. In addition to resisting enclosure, these movements also experimented with new forms of popular sovereignty, animated by a fierce critique of the blindness to inequality that characterizes liberal democracy and the regime of private property rights on which it is founded.[274] New structures of governance were developed in global movements founded on the idea of the people as an egalitarian collective with a mandate to rule in order to bring about social transformation. These experiments reached their highest point with the emergence of what might be termed the social movement party in countries such as Bolivia and Brazil,[275] but the effort to develop egalitarian, non-bureaucratic ways of organizing societies has been a key feature of the Left in recent decades.[276] And, as feminist scholars such as Silvia Federici have documented, contemporary commoning movements crucially include the fight for communal, egalitarian control over material needs linked to social reproduction such as housing, food preparation, child rearing, sex and procreation, and even the reproduction of collective memories.[277]

These radical experiments have exciting implications for the struggle for energy democracy. For example, when the power company came to strip them of their light, the residents of Highland Park took power into their own hands in ways that built on the logic of popular sovereignty developed in global commoning movements. DTE Energy had consistently used political donations (based on those elevated rates) and lobbying to stymie efforts to establish local ownership of clean energy in

Michigan.[278] Now it was taking away the power supplied by dirty coal plants. Faced with this threat, citizens of Highland Park established Soulardarity, a community-based organization that fights for collectively owned streetlights, energy production, and equitable development. Soulardarity not only brings light back to Highland Park, it generates the power to run streetlights from the sun. Soulardarity produces what one observer calls "visionary infrastructure."[279] And it provides local folks with jobs building and maintaining this new solar infrastructure. Through the organization's PowerUP program, the community is able to purchase solar power in bulk and at reasonable rates, and to deploy tens of thousands of dollars' worth of solar infrastructure in the community. But this is not just about transformation of the community's physical infrastructure: it is also about broader social transformation in Highland Park. Soulardarity is a democratic, community-governed membership organization that aims to educate Highland Park residents about what autonomous control of power or energy democracy should look like, and to advocate for community ownership, transparency, and environmental sustainability across the region. Soulardarity advocates for a Community Ownership Power Administration (COPA) as a vital element of a Green New Deal in the United States.[280] Like the Rural Electrification Administration that brought electricity to farms across the country during the New Deal in the 1930s, COPA would provide finance and technical capacity to help local communities across the country make the transition to renewable energy. As Jackson Koeppel of Soulardarity explains, COPA would give municipalities, counties, states, and tribal authorities the legal authority and the funding mechanisms that would allow them to "terminate their contacts with investor-owned utilities, buy back the energy grid to form a public or cooperative utility, and invest in a resilient, renewable system."[281]

In the introduction to their collection of essays on the US movement for energy democracy, Denise Fairchild and Al Weinrub contrast corporate models of decarbonization with the forms of renewable energy being fought for by organizations like Soulardarity. For Fairchild and Weinrub, the former are oriented around the growth imperative of capitalism and are characterized by "a transition to industrial-scale, carbon-free resources without challenging the growth of energy consumption, material consumption, rates of capital accumulation, and concentration of wealth and power in the hands of a few."[282] The centralized nature of power generation and distribution in the era of fossil capitalism has not only led to significant waste, with average losses of 8 to 15 percent of power generated as a result of far-flung transmission lines. It has also helped to make energy invisible and unconscious for many ratepayers, while subjecting others to heightened environmental and health damages, harms that track closely along lines of residential segregation and racialized inequality in the United States. Corporate-owned renewable energy is not likely to challenge this history. By contrast, Fairchild and Weinrub argue, the decentralized renewable energy model fosters community-based renewable energy development that "allows for the new economic and ecologically sound relationships needed to address the current economic and climate crisis."[283] Such decentralization of power, they suggest, is facilitated by the distributed nature of renewable resources: "solar energy, wind, geothermal energy, energy conservation, energy efficiency, energy storage, and demand response systems are resources that can be found in all communities," and consequently provide a foundation for "community-based development of energy resources at the local level through popular initiatives."[284] Fairchild and Weinrub's advocacy of decentralized renewable energy is thus predicated on both the material characteristics of renewable energies and the forms of radical democracy that they hope will facilitate and result from a just transition. For them, transition is about community empowerment rather than simply decarbonization of the grid, as important as the latter may be in the struggle to avoid climate meltdown.

Writing in Fairchild and Weinrub's collection of essays, Cecilia Martinez, director of the Center for Earth, Energy, and Democracy, argues that a just transition to renewable energy will require a shift away from today's energy-as-commodity regime. Martinez suggests that energy democracy requires the construction of an energy commons.[285] What models exist to support the institution and collective governance of such an energy commons, one that diverges radically from today's private property–based regimes of energy control and ownership? For Martinez, the first step is to recognize that energy is not so much a physical object, but rather a "vast array of natural interactions and phenomena for societal use."[286] While energy might derive from natural phenomena all ultimately grounded in the harnessing of solar power, it is inescapably rooted in the social forms and infrastructures developed by humans to exploit solar energy. It is about forms of collective power that are active: in other words, about *commoning* rather than about some pre-given and static *commons*. How might energy be regulated in a more egalitarian manner? Martinez alludes briefly in her essay to the legal structures created over the last few decades to establish a global commons outside the control of any particular nation: founded on centuries-old legal paradigms governing the high seas, today's global commons also includes the atmosphere, Antarctica, and outer space.[287] Martinez also draws on the pioneering economist Elinor Ostrom to argue that diverse cultures around the world and across history have established institutions resembling neither the bourgeois nation-state nor capitalist markets to govern resource systems. Martinez points to indigenous governance models of commoning founded on reciprocity, cooperation, and respect not only between humans but also among humans and the more-than-human world.[288]

What are the conditions for the creation of a new world based on the energy commons? The egalitarian governance systems and legal paradigms discussed by Cecilia Martinez are helpful here. The particular material characteristics of modern renewables such as solar and wind power distinguish them from fossil fuels like coal and oil, but to what extent do these specific material forms, which derive directly from solar power and its effect on atmospheric systems, make for a new, commons-based energy regime that might be termed "Solarity"?[289] What forms of collective, egalitarian governance can the movement for energy democracy draw on as it seeks to challenge the centralized paradigms of energy generation and ownership of the fossil capitalist regime? Do legal paradigms already exist to help community-based organizations like Soulardarity escape from the clutches of fossil capital and adopt solar power on a mass basis? What are the limits of these legal paradigms and what juridical innovations might address these limits? These questions all relate to much broader struggles to establish new, revolutionary forms of popular sovereignty to defend and extend *the commons,* but they have a particular import for the fight for energy democracy. The struggle for a rapid and just energy transition is at the core of broader struggles for an exit from today's trajectory toward social degradation and planetary ecocide. The question of the energy commons is therefore fundamental to the fight for a collective future.

THE TRAGEDY OF THE FOSSIL CAPITALIST UNCOMMONS

Any discussion of the commons inevitably will bring to mind the hugely influential essay on "The Tragedy of the Commons" by the ecologist and geneticist Garrett Hardin. In his 1968 essay, Hardin famously lays out a parable about overgrazing on a village commons in England. Drawing on a pamphlet published by British amateur mathematician William Forster Lloyd in the mid-nineteenth century, Hardin asks readers to "picture a pasture open to all. It is to be expected that each herdsman will try to keep as many cattle as possible on the commons."[290] Therein, according to Hardin, lies the tragedy: "[Each] man is locked into a system that compels him to increase his herd without limit—in a world that is limited. Ruin is the destination toward which all men rush, each pursuing his own best interest in a society that believes in the freedom of the commons. Freedom in a commons brings ruin to all."[291] Hardin builds his gloomy analysis of the ruin of the commons by qualifying it as a tragedy, drawing on a very specific definition of the literary genre derived from the philosopher A. N. Whitehead: "[The] essence of dramatic tragedy is not unhappiness. It resides in the solemnity of the remorseless working of things."[292] Hardin uses the term "tragedy," in other words, to suggest that the despoliation of the commons is not simply an unhappy or sad result of the herdsmen's pursuit of individual gain, but rather that the destruction of the commons is an *inevitable* result of such selfish behavior.

Hardin's brief meditation on the commons has had an outsized impact on modern thought about the environment and society. Indeed, not only is his essay regularly taught in fields as diverse as ecology, economics, political science, and environmental studies, but his argument has also resonated with policymakers at the International Monetary Fund, the World Bank, and myriad conservative think tanks advocating measures such as structural adjustment, austerity, the shrinkage of the welfare state, and the privatization of public resources.[293] Hardin's argument has been used to green-light all kinds of neoliberal policies. As Rob Nixon puts it, "Hardin's pithy essay title and succinct parable have helped vindicate a neoliberal rescue narrative, whereby privatization through enclosure, dispossession, and resource capture is deemed necessary for averting tragedy."[294] Those who employ Hardin's parable to legitimate a new round of enclosures seldom highlight the fact that the essay was intended as a direct attack on core doctrines of neoliberalism. For example, Hardin begins his parable with a broadside against Adam Smith's idea of the "invisible hand of the market," a notion that is fundamental to neoliberal doctrines of laissez-faire capitalism and privatization. Immediately before expounding his parable of the commons, Hardin states that Adam Smith's popularization of the "invisible hand" metaphor "contributed to a dominant tendency of thought that has ever since interfered with positive action based on rational analysis, namely, the tendency to assume that decisions reached individually will, in fact, be the best decisions for an entire society."[295]

But Hardin's influential essay does not simply contradict much neoliberal thinking: it is also shot through with racist, nativist, and eugenicist arguments that are central to contemporary xenophobia and eco-fascism. For Hardin, the widespread acceptance of Smith's doctrine of individual freedom has led to an "inescapable" population problem: "If this assumption is correct it justifies the continuance of our present policy of laissez-faire in reproduction."[296] The core of Hardin's argument is thus a Malthusian diatribe in favor of coercive population control. These were not just academic arguments. Hardin was at the center of the nativist movement in his time, and helped to give that movement legitimacy using his credentials as a scientist and his arguments about the negative impact of immigration on the environment. According to the Southern Poverty Law Center, Hardin served on the board of directors of the enduringly influential Federation for American Immigration Reform (FAIR) and the white-nationalist Social Contract Press.[297] He also co-founded the anti-immigrant Californians for Population Stabilization and the Environmental Fund, whose primary function was to lobby Congress for nativist and isolationist policies. In his essay, Hardin rails against the Universal Declaration of Human Rights and pens lines such as "the freedom to breed is intolerable." It is not too much to say that Hardin was one of the founding figures of contemporary eco-fascism, which has recently inspired mass shootings by white supremacists in sites from El Paso, Texas, to New Zealand.[298] With the aid of massive donations from wealthy elites like the heiress Cordelia Scaife May—whose family fortune was based on banking and oil—FAIR and the anti-immigrant agenda went from being a fringe movement among environmentalists to become the core ideology driving the Trump presidency.[299] Arguments about the fate of the commons are thus central to the contemporary global upsurge of the Far Right and to the eco-fascist arguments cited by both established politicians and neo-Nazi mass murderers.

But the problem with "The Tragedy of the Commons" is not simply that its author was a racist and white supremacist. The parable's argument is also dead wrong. Hardin's parable erroneously assumed that all commons are completely open access, and that there are, in other words, no social means to exclude people who behave selfishly toward the commons.[300] Early pastures were not, as Hardin suggests, free-for-all grazing sites that individuals plundered at the expense of the common good. Instead, societies in various times and places have evolved collective institutions to manage the commons sustainably.[301] The economist Elinor Ostrom, who was awarded the Nobel Prize in 2009 for her research into collective modes of governing the commons, dedicated her life's research to exploring the many modes through which the commons have been managed outside the control of the state and market. Ostrom argued that the tragedy of the commons fascinates scholars because it seems to prove that rational action on the part of individuals may lead to collectively irrational outcomes, particularly in cases in which free riders on common property resources are allowed to continue their destructive behavior.[302] These models tend to assume that users of the commons are incapable of communicating with one another, are consequently unable to change their behavior, and therefore depend on an external power—a kind of Hobbesian Leviathan—to intervene in order to control individual behavior for the collective good.[303] But these models are ridiculous, Ostrom points out, since individuals who use commons can communicate with one another in most instances, and are thus able to devise and modify the rules for use of the commons, and also monitor and enforce those rules themselves. This arrangement is far preferable, Ostrom argues, to rule by an external authority since the idea of efficient centralized control is based on assumptions about the accuracy of that authority's monitoring of commons usage, the reliability of the sanctions it establishes to ensure sustainable collective use, and the supposedly low costs of such centralized administration, none of which tend to be accurate.[304] Against arguments for centralized state control and for privatization, both of which assume that commoners cannot establish and enforce self-governed common-property arrangements, Ostrom documents myriad examples of collective management of common property resources, from farmer-managed irrigation systems, communal forests, inshore fisheries, and grazing and hunting territories.[305] To be successful, participants in such commoning schemes must not only have the capacity to communicate with one another, but must also be able to develop trust in one another, a sense that they share a common future.[306] Also, powerful individuals who stand to gain from dysfunctions in collective governance of the commons cannot be allowed to block efforts by the less powerful to change the rules of the game.

As we shall see, ideas about the potential for collective and egalitarian governance of the commons have become increasingly central to efforts to manage new global commons such as outer space, the internet, and genetic biodiversity. The act of commoning is also pivotal to contemporary radical movements seeking to move beyond the moribund state capitalist bureaucracies of "actually existing socialism" and social democracy in the twentieth century.[307] But what of the terrain that I am calling *the energy commons*? How does Hardin's parable and the counterarguments of critics such as Ostrom hold up in relation to fossil capitalism, a realm where ideas of inevitable tragedy seem, at first glance, to be most appropriate given our current headlong rush toward planetary ecocide? The history of fossil capitalism in the United States—but also in other early petroleum-producing nations, from Romania to Russia[308]—offers a dramatic example of what might be termed "the tragedy of the energy commons." In fact, the history of US petroculture in the twentieth century is largely a story of the need to develop viable forms of collective management of energy resources.

Contrary to most dominant impressions today, the challenge the oil industry faced for much of the last century was not scarcity. Instead, the problem has been oil's overabundance, or, rather, the overproduction of oil. To gain a sense of why this is so, it is imperative to understand the legal arrangements that regulate the exploitation of oil. Since shortly after the beginning of commercial oil production in the US in 1859, the industry has been governed by the so-called *rule of capture*.[309] This legal doctrine is based on English common law, which held that the first person to capture a wild and migratory animal was the rightful owner of that animal. This surprising legal dispensation around natural resources probably gained traction in the United States as a result of settler colonial doctrines of *usufruct*. John Locke's argument in his *Second Treatise on Government*—which justified dispossession of Native Americans by virtue of Europeans' supposed "improvement" of the land through intensive farming—provides the backdrop for American ideas that whoever exploits a resource most effectively has a claim to own that resource. But early judicial opinions concerning rights to oil resources were most explicitly based on the idea that oil was tapped from crevices in rocks through which oil might run long distances from wherever accumulations existed.[310] Oil was therefore analogized to wild animals such as herds of bison, with the implication being that whoever could extract oil legally owned that oil, even if they were pumping it up from under someone else's land. It was only in the 1880s that a more accurate science of reservoir geology developed, one that understood that oil had a stable existence in a stratum of reservoir rock covering a certain area. By this time, the rule of capture had become established legal doctrine. But far from effectively reconciling competing claims to this fugitive resource, the rule of capture generated vicious competition that nearly destroyed fossil capitalism in the United States and, with it, the economy of the country in general.

As Matt Huber explains in *Lifeblood,* an incisive critical history of US fossil capitalism, subsurface oil resources were delegated to private property owners on nonpublic lands.[311] This arrangement, relatively anomalous since in most countries subterranean oil deposits are state property, produced a situation in which oil companies and their "land men" had to negotiate leases with a multitude of property owners, many of whom were impoverished small-scale farmers. As a result, multiple oil outfits often had shared property rights to a *single* oil deposit that stretched across multiple property lines.[312] The law of capture meant that the only reasonable response in the context of this hectic geography of private property owners to a common resource was to pump as much as possible up before your neighbor sucked the oil out from under your own feet. This dynamic is epitomized in the famous "milkshake" scene from Paul Thomas Anderson's *There Will Be Blood* (2007), a film adaptation of Lewis Sinclair's *Oil!* (1927), which was a novelistic exposé of the Southern California petroleum industry. When evangelical preacher Eli Sunday offers to sell the driven oil tycoon Daniel Plainview the rights to oil lying beneath the property owned by one of his parishioners, Plainview first taunts him into confessing that he is a phony and false prophet, and then reveals that he has already taken all of the oil by "capture" from surrounding wells: "I drink your milkshake!" Plainview bellows. He then beats Sunday to death, offering a grim comment on the brutal violence of public life and fossil capitalism in the United States.

While *There Will Be Blood* might seem hyperbolically violent, the brutal competition it allegorizes is right on the mark. The rule of capture produced unbridled competition among lessees to be the first to drill, thereby exploiting the high natural reservoir pressures that would drive the oil to the surface in the phallic gushers that are such potent icons of the oil industry. The result was a forest of derricks, the emblematic image of the US oil industry during the early twentieth century, and a visual testament to the tendency toward overproduction generated by the competitive rush to exploit what was effectively a common resource. But in this overpopulated landscape lay the tragedy: the race to drill meant that the subsurface pressure that produced gushers would quickly be exhausted as competitors bored as many wells as possible, leading to immense waste as massive amounts of oil had to be left in the ground.[313] In addition, the outsize increases in production caused by frenetic exploitation of major new fields led to vertiginous price crashes, major storage losses, and the premature shutting down of higher-cost, but still productive, wells. In sum, the individualistically oriented law of capture helped ensure ruinous overproduction.

This dynamic toward overproduction and waste went from being a regional to a national problem after the discovery of the so-called Black Giant, the massive East Texas Oil Field in October 1930.[314] The oil rush that followed this discovery was kicked off just as demand for oil slackened as a result of the Great

Depression. The result was a crash in the price of oil, which became so cheap that it quickly proved economically ruinous to drill for oil. Although most Americans experienced the Great Depression as a period of scarcity, with masses of people standing in bread lines, the problem in the oil economy—as in other areas such as farming—was actually one of excess, one of overproduction that led to a glut of resources while poverty ensured crippling lack of demand for those resources. In some parts of Texas, oil was selling for as little as two cents per barrel in an industry where $1.00 was considered a break-even cost.[315] Coping with the glut of oil required clamping down on excess production somehow: scarcity had to be enforced, which meant that the independent oil wildcatter had to be controlled. Yet this ran counter not just to the legal arrangements that governed the oil industry, but, even more importantly, to the most fundamental ideologies of American capitalism, including the individualistic frontier ethos of the wildcatter and the laissez-faire ethic of the oil rush.

The Hollywood blockbuster *Giant* (1956) treats the material and social contradictions of the East Texas oil rush in terms of roiling class conflict, with the wealthy landowning Benedict family unsettled by the rise of the venomous lumpen ranch hand Jett Rink following his discovery of oil on a small tract of the massive family ranch. Jett Rink's oil derricks catapult him to dizzying prosperity and political influence, threatening to sunder the Benedict's long-standing relation to the land and their internal unity as a family. The disruptiveness of Jett's rise is represented through his undisguised desire for Leslie Benedict, as well as by his raw racism toward the Latinx residents of the Benedict ranch, whom the family has traditionally treated with a kind of feudal *noblesse oblige*. When Jett tries to seduce the youngest daughter of the Benedict clan, patriarch Bick is ready to defend the family honor with his fists. This open class conflict is derailed when Jett collapses into a drunken stupor, a convenient denouement suggesting that the intoxicating wealth produced by oil inevitably degenerates into moral debauchery.

The reality in Texas was, predictably, far more violent, but also more suggestive of the deep contradictions of fossil capitalism and its mode of regulating the energy commons. In 1931, Oklahoma Governor William Murray declared "martial law" and shut down over three thousand wells in the state. Ross Sterling, a former oil executive who was then governor of Texas, initially shied away from using state power against independent oil producers in this manner. But then rumors began to circulate that a group of landowners and oil producers were preparing to dynamite drilling rigs and pipelines owned by producers of "hot oil"—oil produced in excess of the limits established by the Texas Railroad Commission. Faced with this prospect of violent revolt, the governor declared martial law and sent four thousand armed troops into the East Texas oil fields to clamp down on hot oil, arguing that "there exists an organized and entrenched group of crude petroleum and natural gas producers . . . who are in a state of insurrection against the conservation laws of the State."[316] Less than a year later, however, federal courts declared this measure unconstitutional, again sparking a crisis of overproduction. When FDR assumed the presidency a year later, the big oil companies had begun preparing a bill rather unbelievably calling

for the empowerment of a national "oil dictator" to coordinate
the supply of oil on the national market.[317] While the movement
for federal control of US oil reserves ultimately failed, Congress
in 1935 ratified the Interstate Oil Compact Commission, an
arrangement whereby states would coordinate and control oil
production with federal oversight.[318] In this way, excess production was finally reined in and the oil market achieved stability
once more. As Matt Huber remarks, the state conservation commissions established in 1935 were in fact a kind of oil cartel that
anticipated—and provided the organizing model for—OPEC.

The story of the crisis of oil production during the 1930s
illustrates that mechanisms for governing the resource commons are necessary even in the belly of the fossil capitalist beast.
Indeed, the regulation of the Black Giant is almost a textbook
case in the evolution of institutions for collective management of
the common resources—although these institutions functioned
for the good of Big Oil and the politicians whose pockets they
lined, rather than the good of the body politic in general. Indeed,
the crisis of the Black Giant demonstrated that fossil capitalism
would deploy state power and even violence in order to defang
market-destroying antagonists like independent wildcatters
and other producers of hot oil. The resolution of the crisis of the
Black Giant through collective regulation is a story almost wholly
unknown to the American public, few of whom are aware that
oil production has been carefully controlled by a domestic cartel
for decades. Instead, our collective imaginary of fossil capitalism remains saturated with the mythic figure of a Texas wildcatter like Jett Rink and the ideology of laissez-faire production

that he embodies. Jett's story certainly makes for good drama: like all effective antagonists, he represents a threat to the dominant social order, a force of charismatic anarchy that must ultimately be tamed or obliterated. While it is not exactly a faithful representation of the Interstate Oil Compact's history, there is some truth to this depiction of the tragedy of the fossil capitalist commons. As we have seen, through her research into different empirical examples, Elinor Ostrom has shown that there is a wild diversity of arrangements for the governance of common property resources.[319] And we have seen that to be effective, the rules that control use of the commons must include measures to exclude interlopers who have not agreed to established rules. A wildcatter like Jett Rink is the consummate outsider, a person who, if not as severely exploited by the dominant social order as racialized Mexican characters on the Benedict ranch, nonetheless flouts the rules of the dominant order because of his class exclusion. Yet by the end of the movie *Giant*, Jett has collapsed into an alcoholic stupor and the Benedict family and the landed oil oligarchy that it represents has reassumed its rightful place. Viewers of the film are thus encouraged to feel affection for and perhaps even identify with a certain set of stewards of the land and the fossil capitalist commons. Notably, they are not encouraged to imagine taking steps to assert public control of these resources.

THE SOLAR COMMONS

In 2018, the Tucson Solar Commons was hooked up to the grid, becoming the first Solar Commons project to be built in the

United States.[320] The Tucson project began as a collaboration between a group of local partners in Tucson and anthropologist and legal scholar Kathryn Milun, who helped formulate the idea of the Solar Commons. A Solar Commons Project Research Team that works in close collaboration with the Vermont Law School Energy Clinic joined these partners. The Solar Commons is predicated on the idea that a just transition to renewable energy requires new ownership models to equitably share and localize the wealth of the green energy economy. As Milun and collaborator Dakota Worden put it, "[Sunshine], water, wildlife, the earth's atmosphere and other gifts of nature are 'commons.' They belong to everyone, including future generations, and should be sustainably managed for the long-term well-being of all. In the twenty-first century, local communities are reviving ancient ways and creating new ways to value, protect, and equitably share their commons. Well-managed commons can generate commonwealth benefits which communities can use to alleviate poverty and build resilient and healthy living places for all."[321]

The way all of this works on the ground in Tucson is fairly simple, at least in theory. Community organizations find a Solar Commons donor to supply funding for solar panels, and then create a Solar Commons agreement using the open source legal template created by the Solar Commons Research Team.[322] The host entity (a school, community center, or some other public building) works with a solar installer and the local electricity provider to set up an Interconnection Agreement and install solar panels. Once the panels are hooked up to the grid, they begin to feed power into the grid, generating revenue for neighborhood programs set up to serve as the Solar Commons Trust Beneficiary. In the case of the Tucson Solar Commons, funds from solar power get distributed back to the community in the form of household weatherization, community gardens, and job training programs. The Tucson Solar Commons project also includes a water-harvesting project and a public art project. Created through a collaboration with the University of Arizona's School of Art and College of Architecture, the goal of the public art project is to ensure that communities who are beneficiaries of a Solar Commons trust recognize that they are collective owners of the commonwealth benefits of the sun's energy. The Solar Commons thus helps to establish radical new forms of political subjectivity, offering new material infrastructures grounded in collective governance of the common resource of solar power while also conjuring up forms of egalitarian community based on creating a new world from the old.

Solar Commons designers Kathryn Milun and Matthew Grimley suggest that the collective management of renewable energy can draw inspiration from the community land trust concept in the United States.[323] For activists alarmed by the urban housing crisis in the mid- to late twentieth century, land trusts offered an alternative to private landownership, which they saw as inevitably generating land speculation in cities, mass dispossession, and spiraling economic inequality.[324] In the first publication on community land trusts in 1972, Robert Swann and his associates surveyed examples of land trusts, from Native American tribal lands to Alaskan Native land claims, and concluded that the goal of these various examples was to hold land in trust in perpetuity in order to secure "user rights" to the land through long-term leases. User rights, they argued, allowed people owning leases in a land trust to make and own "improvements" on the land since the land trust concept was concerned not primarily with *common ownership* but with *ownership for the common good*.[325] Milun and Grimley explain that the "key characteristics of trust ownership (institutional prototyping, user rights, individual and collective ownership of the "improvements," and ownership for the common good)" have all been adapted from community land trust to the concept of the community solar trust that structures the Solar Commons.[326]

In addition to drawing on the land trust model, the Solar Commons is inspired by many examples of commoning from the past, from the struggles of medieval English subsistence farms for acknowledgement of their rights by the state that led to the framing of the *Magna Carta*, to the indigenous commons that European settlers from Columbus onward encountered in the Americas, to contemporary struggles for collective social reproduction around the world.[327] These diverse historical examples underline that the commons and the practices of commoning that sustain them are not something that once existed but that have now been gobbled up by capitalist enclosures. While it is true that capitalist enclosures have de-valorized and tended to make practices of commoning relatively marginal, commoners continue to resist the logic of capital accumulation by establishing new forms of working, living, and collective being.[328] For Milun and Grimley, the generativity of the land trust model for the Solar Commons is underlined by its contemporary application in village solar trusts established in the Indian state of Maharashtra.[329] As remote Indian villages that are not hooked into the grid gain access to electricity through solar microgrids, they draw on the village trust ownership model established in the postcolonial era. When land was redistributed from large estates to small farmers after India gained independence in 1947, activists saw that land speculators were swooping in to buy up small plots and restore large monopoly estates. To reverse this trend, rural activists led by Gandhi bought land back from estate owners and placed it in village trust ownership. This ensured that redistributed land would not fall back into the hands of rich landowners. Village trustees could set rules for leasing trust land to local farmers, turning the land into a common good managed according to local rules that increased prosperity and built social trust for the community. Research done by Milun and the Solar Commons research initiative found that the principles of the village land trust are being applied to solar energy in newly electrified tribal villages in India. When solar microgrids are donated with a stipulation that at least half the village trustees be women, Milun and her research collective found that village solar trusts are helping effect social change in places where women have traditionally been barred from owning property and taking leadership roles in village governance.[330]

As these examples suggest, the Solar Commons offers a model for moving beyond the energy-as-commodity paradigm that characterizes both carbon capitalism and dominant approaches to contemporary renewable energy.[331] Indeed, the model of the Solar Commons makes eminent sense as a democratic mode of governing a source of energy that is widely distributed. Solar energy bathes the whole world. Similarly, although some communities might be blessed with greater quantities than others, wind and geothermal energy are resources that can be found in and developed by all communities.[332] In addition, all communities can establish measures for energy conservation, efficiency, storage, and demand response, steps that are key elements of contemporary community-based renewable energy systems. The cutting edge of energy innovation today lies in these so-called distributed energy resources (DERs). To treat energy in this manner, as a distributed commons, is to challenge the centralized model of energy that has historically defined fossil capitalism, with its hierarchical control over and commodification of energy flows. To rework the infrastructure of fossil capitalism is also to radically revamp current power relations. In *Carbon Democracy*, the historian Timothy Mitchell argues that the material nature of fossil fuels plays a fundamental role in the top-down forms of political power that have characterized the fossil capitalist era. From the start, Mitchell explains, "[Oil] production has been based on attempts to evade the demands of an organized labor force."[333] Oil facilitates plutocracy because of the way it flows. Workers found that, because of its liquidity, oil and the infrastructure through which it was produced and distributed was much harder to shut down than coal, the previous dominant fossil fuel. As Mitchell puts it, "[Whereas] the movement of coal tended to follow dendritic networks, with branches at each end but a single main channel, creating potential choke points at several junctures, oil flowed along networks that often had the properties of a grid, like an electricity network, where there is more than one possible path and the flow of energy can switch to avoid blockages or overcome breakdowns. These changes in the way fossil fuel energy was extracted, transported, and used made energy networks less vulnerable to the political claims of those whose labor kept them running."[334] In other words, where miners could effectively shut down coal production by going out on strike, effectively freezing modern civilization, workers in the oil industry found it far harder to take successful action against a slippery commodity like oil. It was not impossible to shut oil production down. But it was, Mitchell argues, far harder to control the industry—indeed, the switch from coal to oil was to a significant degree part of a strategy to defeat militant organized labor in the coal fields.

Oil not only fueled capitalist power as a result of its material nature: it also helped repress and exploit the working class. For radical philosopher George Caffentzis, oil provided energy to automated machinery that has been used for the last two centuries to dominate workers.[335] As the famous assembly line scenes from Charlie Chaplin's film *Modern Times* illustrate, the machine gave capital the ability to dominate workers, subordinating their labor power to the inhumanly fast rhythms of the machine. The globalization of production that has taken place over the last few decades, an innovation which has helped smash organized worker power in the advanced capitalist nations, would not have been possible without computers to coordinate supply and demand and oil-burning cargo ships that transport goods from assembly lines on the other side of the world. The capitalist class also uses machines in an effort to liberate itself from dependency on nature. But these efforts to win autonomy from labor and nature must ultimately fail because they are at bottom self-contradictory: they are always ultimately based on dependence on the labor of other workers and the exploitation of other natural resources.[336] The foremost example of the latter is of course fossil fuels: without the energy derived from coal, oil, or other fossil fuels, energy that always has to be mined or drilled for by workers, machines would no longer function. So at the end of the day the capitalists can never totally free themselves of nor completely dominate workers and nature. Yet for much of the twentieth century, the slipperiness of oil tipped the scales decisively in favor of the capitalist class.

Today, the shift to distributed renewable energy could play a role in reversing the overwhelming hegemony of capital that oil's material nature helped facilitate. But renewable energy will not only be distributed: it will also be digital. In other words, it will be controlled through the sophisticated networks made possible by computers. This became clear in the summer of 2014 in New York City, when demand for power surged as people coped with a heat wave by cranking up their air conditioners. The traditional solution to surging peak demand would be for the utility to build another substation, but this would have cost an estimated $1.2 billion and would have involved a traditional centralized approach to the grid tied to the creation of additional fossil fuel–guzzling infrastructure vulnerable to climate change. Instead of adopting this traditional approach, Con Ed established what has come to be known as the Brooklyn-Queens Demand Management

(BQDM) program.[337] The initiative hinges on conservation measures that save energy (generating so-called nega-watts), as well as power generated from distributed power sources like neighborhood-scale solar and fuel cells. Demand management can include relatively traditional conservation features like efficient light bulbs, but increasingly also involves the creation of digital networks and platforms that allow energy consumers to control their interactions with the grid and with one another.[338] In other words, even though the grid is a social creation, an energy commons, ratepayers at present interact with the grid solely as passive consumers, with no ability to determine how they interact with energy flows. But the increase of distributed energy generation and the mounting stresses on the grid produced by climate change mean that the grid needs to be managed in a far nimbler manner. Instead of a centralized command-and-control structure, which critics such as Amory Lovins have been arguing for years is not only brittle but extremely wasteful, new peer-to-peer technologies allow people to see where their energy is coming from and make choices about how much power they will consume and when they will do so.[339] For Jesse Morris of the Rocky Mountain Institute, decentralized computing solutions like Blockchain, which allow users to engage in secure and decentralized exchanges with one another, are far better ways of managing an increasingly complex grid than a centrally controlled system.[340]

While we're a long way away from visionary calls for the grid to be composed of clusters of microgrids, with control of power managed by transparent access to information about power and peer-to-peer interactions, hints of this future are already visible. In Brooklyn, for instance, a microgrid has been built in which members of a neighborhood solar co-op are able to monitor their energy consumption using a smartphone-based app and trade energy with one another, without the intervention of Con Ed.[341] For Brooklyn Micro Grid subscriber Garry Golden, solar panels, batteries, and digital control of energy are an interwoven package that will make communities more resilient since they can operate autonomously of the grid when necessary, but they are also an essential element in the transition away from fossil fuels: by empowering consumers with knowledge about their electricity use, they will help communities radically reduce the amount of energy they use, an absolutely crucial step if carbon emissions are to be successfully mitigated on the scale necessary to avoid planetary ecocide.[342]

The egalitarian prospects of a decarbonized, distributed, and digital energy commons have been hailed by energy democracy activists, who see these developments as crucial to allowing "working people, low-income communities, and communities of color to take control of energy resources from the energy establishment and use those resources to empower their communities—literally (providing energy), economically, and politically."[343] Struggles within the energy sector may also be seen as indicative of broader shifts in the structure of contemporary capitalism. For Michael Hardt and Antonio Negri, for example, the distinct modes of production of contemporary, digitally based capital are producing new possibilities for social cooperation and transformation. As Hardt and Negri put it,

In the newly dominant forms of production that involve information, codes, knowledge, images, and affect, for example, producers increasingly require a high degree of freedom as well as open access to the common, especially in its social forms, such as communications networks, information banks, and cultural circuits [. . .] all forms of production in decentralized networks, whether or not computer technologies are involved, demand freedom and access to the common. Furthermore the content of what is produced—including ideas, images, and affects—is easily reproduced and thus tends toward being common, strongly resisting all legal and economic efforts to privatize it or bring it under public control. The transition is already in process: contemporary capitalist production by addressing its own needs is opening up the possibility of and creating the bases for a social and economic order grounded in the common.[344]

For Hardt and Negri, in other words, contemporary capitalism produces novel forms of networked cooperation within which the workers themselves generate and control the means of production. Drawing on the tradition of autonomist theory, which suggests that workers rather than capital are the driving force of development, Hardt and Negri argue that capitalism is producing new forms of collective labor that are increasingly autonomous of both capital and the state: a new form of common wealth. Hardt and Negri hold that there is thus a kind of immanent trajectory of cognitive labor that leads it to increasingly break free of the shackles of capitalist domination. We can see similar dynamics on the cutting edge of the energy sector, where the three

Ds—decarbonization, decentralization, and digitalization—are making possible new forms of energy commoning.

But who gains access to the new energy commons? And what is to stop the rich and powerful preying on the commons that communities have laboriously built? As radical geographer David Harvey argues, the forces of reaction can appropriate the commons; indeed, neoliberal capitalism is constantly raiding existing public infrastructure, thereby diminishing the available common wealth.[345] If the commons is less a thing than a social practice—an act of commoning—then establishing and defending such collective, non-commodified social relations is challenging at the best of times. Setting up a solar commons is not so easy, after all: solar arrays are not cheap, and many people lack the capital and the physical space necessary for their installation. Given the extreme inequalities of urban terrain organized around financialized accumulation, organizations fighting for energy democracy in low-income communities and communities of color often must turn to relatively accessible municipal facilities when seeking space for the arrays that power solar co-ops. In addition, less than half of US community solar projects have any participation from low-income households; of these, only about 5 percent of these co-ops include a sizable share of poor folk.[346] Under pressure from the climate justice movement, however, twelve states and the District of Columbia have developed a series of mandates, financial incentives, and pilot programs to help low-income communities access shared solar. These initiatives are transformative; when fully rolled out, they will impact about fifty million households, or 44 percent of the population in the United States.[347] They will bring all of the benefits sought by energy democracy activists, including collective access, economic empowerment, and community control. Gaining access to the empowering benefits of community solar thus requires engagement with and pressure on the state, whether on an urban, state, or federal scale. But it also requires figuring out how to construct energy infrastructures that can exist autonomously from the grid, while also at times pumping power back into that grid in order to access the resulting credits.

In sum, community solar power must deploy a politics that exists "in-against-and-beyond" the state, in the words of energy democracy activist and scholar James Angel.[348] Rather than cultivating imaginaries of complete energy autonomy, advocates of energy democracy seek to intervene in and modify what radical theorist Nicos Poulantzas called the "relation of forces within the state."[349] For Poulantzas, the struggle for socialism consisted in "the spreading, development, reinforcement, coordination, and direction of those diffuse centers of resistance which the masses always possess within the state networks."[350] Poulantzas's arguments about the state's variegated character and the importance of mobilizing around sectors and scales where community power has gained some purchase is particularly relevant in relation to the contemporary energy commons. Poulantzas's injunction about the necessity of struggles over state power reminds us that while distributed renewables might help empower communities like the residents of Highland Park in Detroit, the capacity of such communities to harness renewable power for collective liberation does not flow automatically from the tendencies implicit in distributed energy but rather ultimately depends on political struggle. Although decarbonization of global energy systems is absolutely imperative to avert planetary ecocide, a transition to renewable energy alone will not necessarily heal the forms of galloping inequality and rampant environmental degradation that characterizes contemporary hyper-capitalism. Genuinely publicly controlled systems of distributed energy will help promote regenerative rather than extractive economies only if they are embedded in broader movements for social transformation.[351] The battle for the Solar Commons, with its democratization of energy flows and promotion of new economies based on sustainable economic and ecological relationships, must be won by establishing new models of collective governance and the legal structures to legitimate these models.

DECOLONIZING THE ENERGY COMMONS

In the 1970s, the University of Caracas in Venezuela formed a choral group to tour the Organization of Petroleum Exporting Countries (OPEC) member states, performing folk music from each country. The countries listed on the group's album sleeve include Algeria, Iran, Iraq, Libya, Kuwait, and, of course, Venezuela—nations whose mineral wealth subsequently brought social inequity, political instability, and imperial pillage. The voices of the indigenous people in these countries have been muted if not totally obliterated, their collective cultures and the songs they sang thrown into turmoil, as they were inducted into waged labor and the often-violent tribulations of petroculture. At roughly the same time as the formation of the

OPEC choir, Venezuelan energy minister and OPEC founder Juan Pablo Pérez Alfonzo warned in a prophetic speech of the baleful outcome of the country's oil wealth, saying "Ten years from now, twenty years from now, you will see; oil will bring us ruin. Oil is the Devil's excrement."[352] Alfonzo's warning was uncannily accurate: in subsequent years, Venezuela's oil wealth generated a massive influx of capital, causing the national currency to soar in value, which in turn led to inflation and a slump in manufacturing and other sectors of the nation's economy. The distorting and damaging economic impact of oil in Venezuela became known as the "resource curse," but it is perhaps most powerfully and enduringly characterized by Alfonso's prophetic denunciation of oil as the "devil's excrement."

The energy imaginary in the United States, the center of petro-imperialism, has begun to align with Alfonzo's menacing description of oil. Oil used to be rendered in postwar US popular culture in ecstatic terms, as the kind of secular baptism in miraculous wealth embodied by Jett Rink in *Giant* after his discovery of oil (1956). As the philosophers Antti Salminen and Tere Vadén argue, the large-scale exploitation of oil coincided with the death of God, with oil functioning as a surrogate for divine power and potential: "[Everything] that smacks of being sacred," as they put it, "is burned in the black motor of economic growth."[353] For Salminen and Vadén, fossil fuels are a form of unrecognized and taken-for-granted work, an apparently miraculous force that has an effect that they call "con distancing": fossil fuels "bind the familiar and the close to something unknown, maybe to something nonexistent, to something terrific."[354] Oil's connection of quotidian acts of consumption to the ineffable, to the deep time of eons of solar power concentrated in fossil fuels, tend to remain unconscious because of the unique capacity of this energy source to generate massive, globe-straddling flows of resources. Examples of "con distancing" cited by Salminen and Vadén include the links between the fibers in our clothes, the fillings in our teeth, and the minerals in our phones to environmental disasters and social oppression in distant lands. While those residing in petrocultural sacrifice zones have seldom been granted the affordances of "con distancing," it is increasingly difficult for the beneficiaries of petro-empire to maintain the energy unconscious generated by con distancing. By the time *There Will Be Blood* (2007) was produced, for example, Hollywood had begun representing oil not so much as a blessing than as a column of fire whose infernal and violent denouement is always already present, as inscribed in the film's very title. Today, with the end of easy-to-retrieve oil in the United States and the corresponding rise of extreme forms of extraction such as deepwater drilling, mountaintop removal, and hydrofracking, the energy imaginary focuses correspondingly on landscapes whose wreckage makes them resemble not simply a scene from Dante's *Inferno* but an asteroid-scarred and pitted lunar landscape.

Under such circumstances, the attractive pull of a solar imaginary is easy to understand. In comparison with oil, renewable energies like solar power inevitably acquire heavenly associations, embodying all of the qualities often ascribed to the divine: dazzling light, healing warmth, ubiquitous and infinite goodness. In the nascent genre of solarpunk—a form of speculative writing that refuses the gloomy outlook of much climate fiction (aka Cli-Fi) by attempting to imagine a world after transition to renewable energy—writers tend to be so fascinated by the aesthetic and social possibilities of the abundant energy that streams down onto earth every minute of every day that they imagine that solar infrastructure will be as weightless as light.[355] The coming "absolute distributability" of solar panels for many solarpunk writers portends a world in which they adhere to every surface, generating imagined worlds in which jewel-like solar panels coat and transform the urban fabric into "towering green and silver spires."[356] These giddy images conjure up a hoped-for world that has not simply overcome energy scarcity but has turned energy into a global commons, available to all, with corresponding implications for the overcoming of the forms of energy centralization, social stratification, and imperial violence unleashed in the age of carbon capitalism. They remind us that the call to revive forms of commoning gained prominence among environmental and global justice movements in the 1990s who were struggling against what they regarded as a fresh wave of enclosures unleashed by neoliberal capitalism.[357] The commons for these movements designated various practices, struggles, institutions, and research all dedicated to realizing a noncapitalist future. Indeed, as the political theorists Pierre Dardot and Christian Laval argue, commoning movements hinge on a radical political reversal resulting from the clash between capital and the environment: "[Whereas] the common was hitherto conceived as a great threat to property, which was propagated as the means and reason for living, it is this same institution of private property that has now become the most serious threat to the very possibility of life itself."[358] The tragedy of the fossil capitalist uncommons now threatens to consume all life on Earth. Efforts to imagine a world beyond fossil fuels, to cultivate what has been called "solarity," take up the challenge articulated by Dardot and Laval: how to escape planetary ecocide by bringing into being forms of collective and egalitarian sociality.[359]

This new world is, crucially, made possible by a distributed and democratized energy commons. As energy democracy activist Cecilia Martinez argues, energy transition must end the linked phenomena of environmental degradation and inequality.[360] "Pursuing a sustainable energy future," Martinez holds, "requires moving from the energy-as-commodity regime to 'energy as commons.'"[361] For Martinez, the first step to establish an energy commons is to move beyond a static view of energy as a thing to understand that it is instead a dynamic and collectively achieved relation, an act of commoning that consists of "the transformation of a vast array of natural interactions and phenomena for social use."[362] Martinez argues that a fundamental principle of the energy-as-commons approach is that these natural endowments "should not be owned by, or belong to, any set of peoples, countries, or corporations exclusively. Nor should any one generation assume the right to overexploit or exhaust these resources."[363] Finally, Martinez suggests that "the organizations that extract, refine, transform, and transmit energy should operate as a democratic system."[364] Importantly, Martinez specifies that the energy commons will not be established and maintained without rules and institutions to govern energy-related organizations. As examples of such rules of energy commons governance, Martinez points to the concept of the global commons within international law, which specifies that there are geographical areas outside the reach of nation-states that should be treated as the shared heritage of humanity. Examples of such juridically constituted global commons include the high seas, the atmosphere, Antarctica, and outer space.[365]

Legal efforts to adjudicate problems generated by the last several decades of technological and economic globalization have been guided by a centuries-old tradition of Western legal theory predicated on managing common property on a global scale.[366] This tradition, while certainly not the world's only legal corpus, nevertheless plays a decisive role in governing efforts to establish an energy commons today. For Solar Commons activist and scholar Kathryn Milun, "when you study energy policy, you see that the grid has created contracts or franchise agreements using the vestiges of common property law to establish private and state property."[367] Given this fact, Milun asks, "[What] are the vestiges of common property in our legal system that can be revived and hacked in order to capture the distributed materiality of the sun and its technological potential?"[368] The Solar Commons framework discussed in the previous section is Milun's effort to do precisely this. It is important to note, however, that in her published work, Milun raises potential red flags concerning models of the global commons through her exploration of the colonial genealogy of current legal regimes relating to extra-state domains such as the high seas and outer space. In *The Political Uncommons*, Milun is particularly interested in recent attempts to go beyond the international order of the nation-state to create legal institutions for global commons such as Antarctica, the radio frequency spectrum through which wireless Internet service functions, and even the wealth of genetic material that constitutes global biodiversity. Legal doctrines developed since the 1960s have attempted to manage these extra-state commons for the collective benefit of humanity. Yet the legal doctrines governing the commons should not be seen as a set of value-free, precise tools, Milun argues, but rather as imaginative narratives through which the boundaries of communities are defined and maintained, in the same way that novels and newspapers have been said to help conjure national communities into being.[369]

What imaginary does global commons jurisprudence draw on? The global commons are described in contemporary legal theory as nonstate spaces, or as spaces outside the territorial sovereignty of the nation-state. The legal precedent for the status of the global commons reaches back to ancient Rome, where things such as the air, running water, and the sea were held to be *res communis omnium*, or things that by definition cannot belong to anyone and instead are the common property of all. But there are grave problems with this legal designation. For example, in their book *Common*, political theorists Dardot and Laval point out that the category of *res communis* is an inherently toothless one since it is pre-juridical: a thing such as the ocean is held to be common by virtue of its vast extent, abundance, and difficulty to capture. *Res communis*, they argue, cannot constitute a fully legal category since its status as a common good results from the essential nature that is ascribed to it, while everything categorized as public is constituted only in relation to a dispute or trial of some kind.[370] There is, in other words, no truly pre- or extra-social space. Moreover, there is a slippage, Dardot and Laval argue, between this legal category of *res communis* and the related category of *res nullius*, which designates those things that, being claimed by no one, belong to those who first lay claim to them.[371] The upshot is that things designated as the collective heritage of mankind have tended to become the property of the first person or state strong enough to jab a flag into them. In this context it is worth recalling the *rule of capture* doctrine, according to which fugitive things such as pheasants, bison, or oil became the property of whoever successfully captured them first. As we have seen, this doctrine essentially became a rubber stamp for practices of accumulation by dispossession, leading to the tragedy of the fossil capitalist uncommons.

It should come as little surprise that these immensely influential legal doctrines for staking claims over territory outside the boundaries of European states have a strong colonial provenance. In *The Political Uncommons*, Kathryn Milun agrees with Dardot and Laval in their assessment of the slippage between *res communis* and *res nullius*; she goes a step further by linking the latter category to *terra nullius*, or vacant land.[372] Milun shows how the doctrine of *terra nullius*, or "unclaimed land," and related legal categories such as *vacuum Domicilium*, or "empty home," were used by European colonial states to dispossess indigenous people of their land in the eighteenth and nineteenth centuries.[373] These legal categories were essentially imperial discourses through which land was imagined as empty, undeveloped and hence unclaimed and available for the taking. For Milun, contemporary legal designations carry these forms of colonial sovereignty into the heart of struggles over spaces outside the territorial sovereignty of the nation-state. Rather than protecting the global commons as *res communis* or the property of everyone, international law treats the global commons as *res nullius*, domains that cannot be governed and therefore belong to no one—at least until they are captured by someone. In doing so, the international law of the global commons works as a kind of technology of accumulation by dispossession. This sets the earth's nonterritorial spaces up not just for privatization but also for ruin, since Western legal theory treats the commons both as natural spaces accessible for use, exploitation, and dispossession by modern technological innovations, *and* as a sink where the polluting "externalities" generated by the industrialized nations can be dumped.

The discourses of colonial sovereignty embedded in legal doctrines through which global commons are structured have important ramifications for how we imagine and structure the energy commons of the future. The distributed and democratized aspects of renewables that give rise to an imagined energy commons are only one avatar of the energy transition: the centralized, capitalist, and imperialist model of energy that has traditionally defined fossil capitalist culture is very much alive as the world struggles to transition to renewable energy. Solar power and wind power may be more distributed and hence more materially amendable to grassroots popular control, but they can nevertheless be subordinated to the centralizing whims of plutocrats. Large wind farms and big solar plantations are, at present, a product of concentrated financial and political power.[374] Such large-scale forms of renewable energy are very rarely established through the democratic action of communities; instead, they tend to reflect today's overwhelming concentration of economic and political power in the hands of large corporations. And they in turn generate benefits for those corporations and their shareholders instead of for the communities in and near which they are situated. Energy democracy activists must be wary lest the forms of colonial sovereignty that are reappearing in global commons jurisprudence reassert themselves in relation to the energy commons as well.

This worrying potential is most palpable in relation to a seldom-discussed but absolutely crucial component of renewable energy: power density.[375] This term refers to the energy flow that can be harnessed from a specific place or system. It is measured using units such as horsepower per cubic inch, watts per kilogram, and watts per square meter (W/m^2). The latter unit is by far the most salient when it comes to renewable energy sources like solar and wind power since it indicates how much energy can be derived from a given piece of land. If a source has low power density, it will require large amounts of space in order to generate the substantial quantities of energy the world currently demands. This is important because we are transitioning from a world oriented around the extremely high power density of fossil fuels to one based on the relatively low power density of renewable energies. For example, wind energy generates, on average, 1.2 W/m^2; solar photovoltaic generates 6.7 W/m^2. Both sources are also intermittent, meaning that they require significant storage resources in order to be functional for societies that demand power 24/7. Compare this relatively meager power density with that of fossil fuels: a marginal natural gas well, producing sixty thousand cubic feet per day, has a power density up to 1,000 W/m^2. Oil and gas wells in other words have power densities up to three orders of magnitude larger that renewables.[376] Solar panel technology is likely to improve power density to a certain extent, but renewables will never generate anything even remotely approaching the power density of fossil fuels. The low power density of renewables means that more land, steel, and ultra-long transmission lines must be used to generate equivalent amounts of power as nonrenewables. Way more. A recent study estimated that solar and wind power need around forty to fifty times more space than coal and ninety to one hundred times more space than gas.[377] Only one half of 1 percent of US landmass is currently devoted to the current, fossil fuel–dominated energy system. But a fully renewable system is likely to occupy one hundred times more land.[378] In an energy-hungry country such as the United States, this means that a region such as the northeast would have to devote 10 percent of its land to energy generation.[379] Whose land will get occupied to generate the renewable energy needed to run capitalist societies at current levels of consumption?

The transition to renewable energy is likely to intensify existing forms of extractivism and neocolonialism unless inequalities of control over territorial rights, technology, and capital are directly addressed in order to construct a genuinely egalitarian energy commons. While receiving less attention than dispossession resulting from pipeline construction, the requisitioning of indigenous land in the name of climate change mitigation and a putative "greening" of the economy, an iniquitous and often violent process also known as "green grabbing," is all too real in many parts of the world today and certain to intensify as the shift to renewable energy gathers pace. For example, in the Isthmus of the Tehuantepec region of Oaxaca, Mexico, a region that the World Bank describes as having "the best wind resources on earth," renewable energy companies based in the Global North have dispossessed indigenous communities in the course of a "wind rush."[380] After a 2003 USAID report publicized the region's superlative wind resources, transnational European and US companies like Iberdrola, Gamesa, Vestas, and Clipper Windpower moved quickly into the region, known locally as the Istmo, building twenty-seven wind parks and more than 1,800 wind turbines by 2016.[381] The Istmo's wind resources would not have been so attractive were it not for the neoliberal transformation of Mexico's economy: Mexico's Electricity Law of 1992 and the 1994 passage of NAFTA meant that the majority of the wind parks in the region could be established on a "self-supply" basis, meaning that the electricity they produce is private. The power generated by a common good like the wind, in other words, is reserved exclusively for the investors and owners of the wind parks, who transport the electricity generated in the rural Istmo to industrial facilities in Mexico, to Guatemala, and even to the United States. Local elites worked with the big wind companies to market the wind parks to the Istmo's predominantly indigenous communities with promises of economic development.

Wind power, it was said, would be a solution to poverty and unemployment. Yet only a few landowners have benefited from the power generated by the wind turbines, which have brought temporary work and token social uplift projects while catalyzing rising income inequality. Residents of indigenous villages such as La Ventosa now find themselves completely surrounded by wind turbines, with few material benefits as a result and widespread health and environmental concerns.[382] As the character of green grabbing in the Istmo has become clear, indigenous communities have resisted this new form of extractivism with increasing militancy. The residents of La Ventosa, for example, have been struggling for greater inclusion in wind energy profit shares as well as access to civil infrastructure such as sewers, roads, and improved health care facilities.[383] Communities that have rejected wind parks have been subjected to ideological warfare, including media slander and PR campaigns, as well as far more direct forms of violence, including assaults, death threats, and assassinations. The battle over the wind commons in the Isthmus of Tehuantepec should be seen as a front in what Zapatista Subcomandante Marcos called "the Fourth World War," a globally diffused series of low- and high-intensity conflicts through which transnational corporations and complicit state elites seek to extract the natural resources of the world—nonhuman and human—and feed them into the insatiable maw of capital.[384]

While it is no exaggeration to say that the fate of the planet hangs in the balance, the struggle for a decolonized energy commons is not completely quixotic. In the Istmo, for example, local organizations led by the Assembly in Defense of the Land and

Territory of the Indigenous People in the Isthmus of Tehuantepec convened a forum in 2009 organized around the themes of "indigenous communities, self-determination, and energy sovereignty."[385] This forum not only represented one of the first instances in the world in which the concept of "energy sovereignty" was used, but also opened a space to discuss community-based wind farms as an alternative to the privatized parks growing throughout the Istmo. After the forum, the indigenous commune of Ixtepec, a Zapotec community that has decided to maintain the communal ownership and management of common land and resources, approached the nonprofit Yansa Group with the goal of establishing a communal wind farm.[386] While the proposed project would be similar to private wind farms in terms of scale (the number of wind turbines and installed generation capacity) and the amount of investment required, the ownership and revenue distribution would all be markedly different inasmuch as they would benefit the entire community rather than a small number of shareholders. In addition, decision-making processes for the Ixtepec wind farm would be based on active participation of the community through existing communal institutions such as the Assembly and peasant organizations, as well as new spaces of decision-making like forums of women and youth. The energy sovereignty initiative in Ixtepec shares values of cooperation and reciprocity with indigenous governance models elsewhere; as Cecilia Martinez suggests, these values—antithetical to the norms of commodification, markets, and material abundance that govern our current energy system—offer inspiration for energy transition.[387] But the forces fighting against these alternative values are fierce. The community wind project in Ixtepec has been held up by the Mexican Federal Electricity Commission, which ruled in 2012 that the Yansa-Ixtepec collaboration does not constitute a legal entity recognized by the state. The struggle to build the Ixtepec community wind project continues today, offering both a resource of hope for struggles to establish socially just, decolonial models of the energy commons as well as a sobering lesson about the power of green-grabbing interests arrayed against such struggles for energy sovereignty.

CHAPTER FIVE: PUBLIC POWER

A hard rain was hammering down, flooding New York City streets, as I headed to a hearing on the crisis of the city's power supply in a school in the borough of Queens. Although the monsoon-like storm made it hard to get around without getting soaked, the deluge was very welcome on that muggy summer evening: it finally broke the sweltering heat, which the city had been enduring for the last few weeks. As is always the case these days, the weather was political. In the days preceding the meeting, the city's main power utility, Consolidated Edison (Con Ed for short), had cut power to fifty thousand people in some of the city's poorest neighborhoods in order to avert a wider blackout. Areas of New York where power was cut off, like Canarsie and Flatlands, have some of the highest percentages of residents at risk for heat-related deaths. This risk of mortality results from the histories of segregation and racism to which these communities of color have been subjected: their residents are poorer and less able to afford the high bills that come with running air-conditioning frequently, and their neighborhoods have less cooling green space than more affluent, whiter parts of the city. To make matters worse, Con Ed gave community residents little

warning of the power outages, claiming shortly before the blackouts that the utility was on top of things as the city sweated under
the extreme heat. When they were imposed on poor and heat-
vulnerable parts of the city to maintain power in wealthier
neighborhoods, the power outages ignited public outrage. As
Flatlands resident Stanley Henriques put it, "Why are we the
ones that have to suffer for the bigger power grid? We pay just
like everyone else."[388] Henriques had a point: he and his neighbors were being made to pay for mistakes made by others. Con
Ed had received $350 million to upgrade the relay protection sys-
tems that were blamed for the outages several years ago but ultimately failed to implement these modernization plans. After the
outages, Jumaane Williams, the New York City Public Advocate
and a former councilmember representing many of the neighborhoods struck by the power cuts, promised to hold Con Ed to
account, while Mayor Bill DeBlasio suggested that the city should
either find another electricity provider or take over Con Ed.[389]

The anger was palpable at the community meeting in Astoria
on that rainy summer night. The meeting was organized by a
NYC councilmember to allow the public to weigh in on proposed
hikes in customer payments for electricity and gas in the city. Six
months earlier, in February 2019, Con Ed had submitted a proposal to the New York State Public Service Commission, the body
tasked with regulating the utility, for an 8.6 percent increase
in electric delivery rates and a 14.5 percent increase in natural
gas service. Testimony at the meeting revealed the stark injustices inherent in the city's for-profit power system. New Yorkers
already pay the second highest utility rates in the country, which

is nearly double the national average, for service whose unreliability the recent outages had underlined. Fixed utility rates are a form of regressive taxation since the poor and the rich pay exactly the same rate, and higher rates and energy consumption in general make money for the utility, disincentivizing energy efficiency and solar ownership. As the organizing councilmember pointed out in his introduction to the meeting, Con Ed makes over $1 billion in profits each year, with annual dividends to shareholders rising 79 percent over the last two decades. Yet despite supposedly functioning in the public interest, the utility threatens to shut off power and gas to millions of people each year; between 2010 and 2018 the company terminated service for over 650,000 customers.[390] Unlike water, which is provided by the city to all residents as a right, electricity and gas are available only to those who can afford them. To avoid freezing in the dark, utility customers in New York City turn in large numbers to "payday" loans that come with exorbitant additional fees: paying arrears to utility companies is the largest single reason people take out such extortionate loans.[391] Jacking up rates would inevitably plunge increasing numbers of New Yorkers already pinched by spiraling rents and the generally high cost of living in the city into energy poverty. To make matters worse, Con Ed has a long history of neglecting the grid, refusing to fix problems until they lead to deadly "accidents."[392] To take just one example, instead of finding and fixing gas leaks, Con Ed deals with such leaks by drilling holes in manhole covers, letting methane belch into the atmosphere (thereby exacerbating climate change), while allowing snow and water to drip down and corrode pipes, causing them to smoke or catch fire.[393] Despite this atrocious record, the utility was now requesting another rate hike. What were its plans for the additional revenue generated by higher rates?

Con Ed president Tim Cawley explained the corporation's request by saying, "Our proposal will build on the progress we have made putting tools in the hands of our customers to help them manage their energy usage."[394] Cawley's statement presumably refers to Con Ed's plan to spend $989 million installing smart meters in homes and businesses, a plan that the CEO spins as a form of energy democracy. Smart meters theoretically provide consumers with more information about their energy usage, thereby helping promote conservation, energy efficiency, and consumer awareness that would lower power bills. As Con Ed's CEO suggests, smart meters could be used to implement time-of-use pricing schemes, allowing customers to save money by using power less frequently during times of peak demand, but the utility has not yet used its smart meters to such an end. Although they are not performing as advertised, smart meters are undeniably useful to utilities like Con Ed that wish to track their customers more closely—and potentially disconnect them the moment they fall into arrears. The meters also allow the utility to monetize data gathered by the smart meters by selling this information to other companies. As concerns grow about surveillance capitalism, the wisdom of having an insecure wireless device that is potentially able to monitor all other devices in one's home is in increasing doubt.[395]

In addition to expanding its smart meter program, Con Ed announced that it plans to use additional revenue from rate hikes to install new fossil gas infrastructure, including upgrading a liquid natural gas plant located in Astoria. At the public hearing the implications of these plans became starkly apparent when a young woman stood in front of the microphone in the school auditorium where testimony was being delivered. With a toddler on her hip, she said that she lives a few blocks away in what is known as Asthma Alley. The name, she explained, derives from the high rates of the disease suffered by children as a result of the pollution generated by power plants in the neighborhood. The child she was holding, she said, is among those afflicted. More than 50 percent of New York City's electricity comes from Astoria-based power plants like the Ravenswood Generating Station. For years these plants have burned millions of gallons of the dirtiest fuel available, generating air pollutants like nitrogen oxides, sulfur dioxide, carbon dioxide, methane, and particulate matter. Con Ed and the other private companies that operate these plants burn dirty oil for the simple reason that it is cheaper than cleaner oil, and because using other oil would require expensive upgrading of plants. Driven by the desire to make outsize profits, the corporations that run these power plants have chosen to keep using the most polluting fuel possible. But burning dirty oil has a direct impact on public health. According to the Department of Health, Astoria and Long Island City suffer from worse air pollution than the rest of the borough and the city.[396] The Environment New York Research and Policy Center found in 2014 that the Ravenswood Generating Station was the largest carbon polluter in the state.[397]

The health challenges affecting low-income communities in Queens are symptomatic of broader class- and race-based inequalities related to power generation in NYC. Each year, more New Yorkers are killed by pollution from electricity generation than in any other major city in America. According to a 2013 study, NYC has more than double the number of premature mortalities of any other large metropolitan area in the United States.[398] The majority of the oldest and dirtiest plants are concentrated in low-income communities of color.[399] Stinging public criticism of dirty plants like Ravenswood has had some impact: in October 2019, the state Public Service Commission announced that a giant 350-megawatt (MW) set of batteries—enough to power 250,000 homes for eight hours—would be set up on the grounds of Ravenswood, allowing the city to shut down some of the dirty fossil-fueled "peaker" plants that crank up on hot summer days when demand for energy to power air-conditioning is highest.[400] But Con Ed's plans to expand gas infrastructure run directly in the opposite direction of these reforms. While fossil gas is far cleaner than the oil burned in Ravenswood, it is still responsible for methane emissions that contribute significantly to global warming.[401] By ramping up fossil gas in New York, Con Ed is signaling that it does not intend to abide by the diminished emissions targets mandated in landmark legislation like the Climate Leadership and Community Protection Act, which was passed in 2019 after years of agitation by a large and diverse group of social movements.

After stinging testimony from the likes of the young mother about the detrimental impact of Con Ed's policies, the sentiment in the school auditorium in Astoria on that muggy summer night was clear: Con Ed is failing the people of New York. It delivers bad service, high prices, and failing infrastructure. Con Ed doesn't deserve a raise. In the words of Brent O'Leary from a local civic association, "[We] have to say *no* to rate hikes, *no* to monopolies, and *yes* to public government delivering public services we need so they can be democratic, accountable, and transparent." But what are the alternatives to the investor-owned utility, now dominant not just in New York, but around the United States for nearly a century? A number of campaigns are afoot in New York City to challenge the monopoly utility. Among them are efforts to establish community-based solar cooperatives similar to those described in the previous chapter. Like the efforts of Soulardarity in Detroit, these solar co-ops often emerge from longstanding environmental justice organizations. One such example is Uprose in Brooklyn.[402] But, as is the case with Soulardarity, Uprose's Sunset Park Solar Co-op is not wholly autonomous from the larger grid, which remains in the hands of Con Ed. In addition, the number of people served by such community-based solar co-ops is still miniscule. Indeed, only 1 percent of the households of New York City are powered by solar, whether through co-ops or through individual solar installations.[403]

While supportive of such efforts to establish community-based solar co-ops, New York City activists have also been engaged in a campaign over the last year for a wholesale transformation of the grid. The Public Power NYC campaign took off from opposition to the proposed Con Ed rate hike but over the course of 2019 built to much more ambitious ends, culminating in the demand for "renewable energy as a human right."[404] Public Power NYC activists offered two potential paths to achieve their transformational demands. On the one hand, they suggested that the city should simply take over the utility in a process that is known as municipalization. But, given the legal and financial challenges associated with municipalization, such a process is inescapably a lengthy one. Efforts to municipalize also would surely face stiff resistance from Con Ed; as a nonprofit devoted to the issue puts it in its guide to municipalization, "The incumbent utility will wage a major public relations battle to stop the community from forming a public power utility."[405] This does not mean municipalization is impossible: campaigns for municipalization are under way in cities like Chicago and San Francisco, and public or municipal utilities and rural electric cooperatives serve close to 30 percent of all electricity customers in the nation.[406] Yet the deep pockets and political ties of Con Ed undoubtedly make municipalization in NYC a formidable challenge.

Another strategy advanced by Public Power advocates involves an end-run around Con Ed through expansion of the New York Power Authority (NYPA). Dating back to the New Deal era, NYPA is the largest state-owned public utility in the United States. It currently produces about 25 percent of the state's energy, with a strong track record of delivering reliable low-cost power to public housing residents, the Metropolitan Transit Authority, and universities such as CUNY and SUNY. If NYPA were expanded, it could provide affordable financing for the transition to renewable energy and for energy efficiency upgrades, without having to generate the outsize profits for shareholders that shape the behavior of a corporation like Con Ed. But NYPA as presently constituted is constrained in its abilities: its charter prevents it from servicing new customers or building its own renewable energy projects. It can purchase clean energy from other providers but cannot own such energy. For it to effectively replace Con Ed, NYPA's charter would have to be amended, a move that would of course face significant political obstacles. Public Power NYC activists told me that they generally favor the more democratic structure that could be created through municipalization but recognize that NYPA already has a lot of funding and capital investment that make it a viable alternative to Con Ed.[407] On the other hand, the municipalization law gives the city a right to take over the incumbent utility through a ballot initiative, a popular vote that Public Power NYC activists were increasingly feeling they might win as a result of mounting public outrage following the power outages during the summer of 2019.

In sum, both possible options—outright municipalization of Con Ed or expansion of NYPA—have significant drawbacks and would necessitate intense political struggle. As anger over the Con Ed power shutoffs during the summer heat wave mounted, Public Power NYC activists found their campaign garnering a surprising amount of popular sympathy but were still unsure about which strategy for leveraging this somewhat unexpected political traction would make the most sense.[408] Their decisions were informed by the campaigns of other public power groups across the United States and by a longer history of social movements fighting not simply to transition from fossil fuels to renewable power but for energy democracy. What kinds of strategies had activists in the past and in other cities, states, and countries adopted, and how might public power activists in New York City learn from the successes and defeats of these campaigns? What exactly do activists mean when they campaign for public power? What forms of governance, in other words, can ensure that the provision of renewable energy doesn't assume the kind of top down, unjust character of a so-called public utility like Con Ed? What are the implications for how activists for energy democracy should think about and interact with the state on various scales?

In this chapter, I explore the now-famous energy transition in Germany, unpacking the policy innovations that made the country's shift to solar- and wind-powered renewable energy possible, assessing the reasons for the recent setbacks in the German energy transition, and analyzing the response of social movements to these setbacks. Struggles for energy transition in Germany offer useful lessons for current and future struggles for public power in other parts of the world, including New York City. Granted, energy systems and the cultures within which they are embedded differ markedly, and no situation is exactly the same. In Chicago, for example, where an intense campaign for public power is unfolding, the city's contract with the incumbent utility is expiring, generating an important opportunity to revisit the status quo. What to do following the expiration of Con Ed's contract with NYC, by contrast, has thus far never become a political issue. Nonetheless, there is much to be learned from comparative analysis of fights for energy democracy. The transition to renewable energy needs to be speeded up and scaled up. As we have seen, despite years of promises about market-led growth of renewable power, the battle to move the world off of fossil fuels is not being won. According to the US Energy Information Administration (EIA), wind, solar and other non-hydropower renewables provided a scant 9 percent of total utility-scale generation in the country in 2018.[409] It projects that the proportion of power generated by modern renewables will diminish in 2020. As we will see, the energy transition is also stalling in the European Union. These stark statistics underline the fact that the current model of ownership, one that drives the growth of fossil fuels and extractivism, must be abandoned. The struggle for democratic control and social ownership of energy must be redoubled. As scholars such as Andrew Cumbers have argued, the fight for energy transition offers the Left a unique opportunity to reclaim ideas of democracy and liberty from the Right, although these ideals must be linked to the common good rather than corrosive individual freedoms. [410] Fresh strategies for winning public power—in both senses of the term— need to be devised, and new democratic, decentralized, and participatory forms of governance must be developed to ensure that public power genuinely lives up to its name.

THE (RADICAL) POLITICAL CONTEXT OF THE GERMAN ENERGY TRANSITION

Germany has become well known around the world for its *energiewende*, the country's planned transition to a low-carbon, nuclear-free energy system. Renewables generate roughly 50 percent of electricity in the country, although during a week in spring 2019 they generated nearly 65 percent of Germany's power, with wind turbines alone responsible for nearly 50 percent of the country's energy.[411] The speed of the country's transition to clean power is truly impressive: as recently as the early 2000s, the country got less than 10 percent of its power from renewables. Germany's abrupt decision to shut down its nuclear power plants after the disaster at Japan's Fukushima plant in 2011 drew considerable international attention for its boldness and determination to shift the country to renewables, although, as we will see, the energy transition had been underway for many years prior to this. Among large industrial nations, Germany is a clear leader in the energy transition and environmental politics in general—demonstrating that you can power an economy that pumps out steel and BMWs with renewables. Nonetheless, it is not entirely unique. Other European nations such as Austria, Belgium, Denmark, and Italy have nuclear phaseout plans, and other countries have similar shares of wind and solar in their grids.[412] What makes Germany truly distinctive is the central role played by grassroots struggles for energy democracy in the nation's transition to renewables.[413] Indeed, the term *energiewende* was coined in the 1970s, when a rural community in a conservative part of the country rose up in opposition to government plans to "develop" the area through the construction of a nuclear power plant and the attraction of new industries to buy all the new power.[414] Against the imposition of social and environmental transformation that would benefit big power utilities and industrial interests while leaving locals to shoulder the risks of nuclear power, the community argued for greater democracy in the energy sector and for a prominent role for citizens in planning the country's energy future.

Surprisingly, the demands for greater community control over energy in Germany were to a certain extent inspired by regulatory reform in the United States. In 1977, newly elected President Jimmy Carter delivered a televised fireside chat addressing the nation's energy crisis, which at the time had been unfolding for nearly a decade. Wearing a beige cardigan, Carter exhorted his fellow citizens to turn down their thermostats, "live thriftily," and make US society more energy efficient. Solving the energy crisis was, Carter famously argued, the "moral equivalent of war." As a result of significant cajoling from the Carter administration, Congress subsequently passed an omnibus National Energy Act in 1978 that contained what at the time was a relatively unnoticed clause compelling the nation's utilities to buy and move to market electricity produced by any facility with an output of less than 80 MW.[415] By 1983, when the Supreme Court confirmed the legality of the Public Utilities Regulatory Policies Act (or PURPA for short), states like California had masses of small energy producers rigging up wind turbines and solar plants whose power the big utilities were now forced to buy. From 1981 to 1990, generation from California's wind sector leapt from 10 MW to 1,700 MW, making the state home to 85 percent of the world's capacity of wind-powered electricity generation.[416] Ironically, this expanding market provided an opening for Danish wind turbine manufacturers, whose heavy-duty designs turned out to be much more efficient than the aeronautics-based designs of US firms. Like the United States, Denmark sought to increase its energy independence following the crises of the 1970s, but in the case of Denmark a strong domestic antinuke movement helped ensure that government plans for local alternatives to imported oil slanted toward renewable energy rather than nuclear power plants.[417] Since the large wind turbines being produced in the mid-1980s by manufacturers in Denmark were too expensive for individuals to purchase, groups of people there began investing jointly in wind turbines through local wind cooperatives. As in the United States, legislation mandated that excess electricity generated by these co-ops could be fed back in and sold back to the grid. The result was a thriving wind turbine industry driven by new, decentralized forms of energy democracy.

Environmental and antinuclear movements in West Germany were just as militant, if not more, than those in Denmark during the 1980s. In fact, social movements in the two countries should be seen as part of a pan-European upsurge of protests against the glaring inequalities and injustices of industrial capitalism during and after 1968.[418] As the powers of corporations and bureaucratic governments expanded during the post-1945 era, popular control over increasingly large and significant swaths of social life had been progressively eroded.[419] Job security was remorselessly undermined, family life destroyed, and environmental conditions deteriorated, all while the commodity form subsumed previously autonomous realms of social life. Although there was a notable break between the largely university-based New Left movements of the 1960s and the more radical groups that came out of the ferment of the 1970s, opposition to the degradation of social democracy in the different decades was characterized by a number of similar characteristics, according to the historian George Katsiaficas, including "antiauthoritarianism, independence from existing political parties, decentralized organizational forms, emphasis on direct action, and a combination of culture and politics as a means for the creation of a new person and new forms of living through the transformation of everyday life."[420]

West Germany was particularly ripe for revolt. The country's post-1945 "economic miracle" relied on capital-intensive industries like steel and automobile manufacturing. All this depended on large amounts of cheap power. The German nuclear industry consequently became a sacrosanct symbol of national success for the political mainstream. By the 1970s, however, a large segment of the public began to revolt against this model of development and the energetic and political dispensation that supported it. In February 1975, several hundred protesters in the rural town of Wyhl occupying the construction site of a nuclear plant were brutally dispersed by police.[421] A few days later, twenty-eight thousand protesters descended on the construction site, building a protest encampment that became a global symbol for peace and antinuclear movements. The corrupt character of the government's nuclear policies became clear when revelations about the revolving door between the provincial government and the nuclear industry were made public soon after the establishment of the protest camp. The repression of political dissent by state police forces staffed by significant numbers of former Nazis helped galvanize German activists, who indicted an antidemocratic political system. As we have already seen, the notion of the *energiewende* emerged from these direct-action protests in Wyhl, where activists fought not just for the end of nuclear power but also for a transformation in the kleptocratic systems of governance that perpetuated it. When local farmers mobilized against a proposed nuclear waste dump site at Gorleben in the eastern part of the country in 1977, similar events unfolded: a protest camp that came to be known as the Free Republic of Wendland was established, attracting a diverse group of ecologists, feminists, peace activists, farmers, alienated youth and students. Savage police repression shifted broader public opinion against both the state and the nuclear industry. The evolving and increasingly militant character of the movement became clear in 1981 at a nuclear plant construction site in Brokdorf: when ten thousand police arrived to clear protesters from the site, the over-ten-thousand protesters who had converged on the site fought the police off with rocks, sticks, and Molotov cocktails.[422]

Resistance to nuclear power was part of a far broader group of extra-parliamentary struggles for the cultural and political transformation in Germany that came to be known collectively as the Autonomen (or autonomous movement). With strong roots in feminist politics and urban squatter struggles for housing and cultural spaces, the Autonomen were autonomous not just from established political parties but also from the traditional Left. Drawing on a feminist "politics of the first person," the German Autonomen developed forms of self-managed consensus rather than relying on powerful individuals or hierarchical organizations to dictate their collective goals.[423] The centralized, technocratic, and potentially deadly character of nuclear power represented the absolute antithesis of the antiauthoritarian, anti-patriarchal, anti-capitalist, and anti-imperialist values that were central to the various groups that came to form the movement. These autonomous groupings existed in a tense but fertile relation with elements of the Left that organized within the existing political system. Although it has often been seen as the principal incarnation of the German Left, the Green Party in fact grew out of the grassroots mobilizations, or *Burgerinitiativen*, against nuclear power, sexism, and capitalism of the 1970s. Indeed, when it was founded in late 1979, the Green

Party was intended to give political expression to the demands of the emergent antinuclear, peace, feminist, and environmental movements that had been growing in Germany throughout the decade. The Green Party thus derived much of its dynamism from the militant resistance of the country's various extra-parliamentary groups. Importantly, and as a result, the Greens were devoted not just to the passage of specific environmental policies but also to transformation of the undemocratic political culture that had empowered the nation's nuclear industry, as well as to overcoming the oppressive cultural attitudes that trampled on the rights of women, gays, and young people to free expression.[424]

The renewable energy policies that the Greens were instrumental in establishing thus need to be seen as emerging from the context of radical political ferment produced by the Autonomen. Germany's remarkably rapid transition to renewable power can only be understood in light of the broader political mobilization effected by the nation's militant extra-parliamentary groups. For these groups, the country's energy system was a clear expression of the antidemocratic nature of politics in general. Pro-renewable power campaigners in the 1980s in fact talked about the German utility system as a set of "power dictatorships."[425]

VICTORIES AND CHALLENGES IN THE ENERGY TRANSITION

Like that in the United States, Germany's grid is organized around a series of publicly regulated regional power utilities that have monopoly rights to generate and distribute electricity.[426] The fact that this centralized system was created during the Nazi era helped underline the authoritarian political calculus that underlay the grid as protesters mobilized against the nuclear industry in the 1970s and 1980s. Intransigent resistance on the part of Germany's big utilities to clean and decentralized renewable electricity further dramatized the dysfunctional character of the country's power system. The monopolistic character of the system also underlined the bankruptcy of free market dogma since unchecked market forces clearly had not produced a proliferation of agile and competitive energy firms but instead had cemented the power of a small number of big utilities with fixed assets sunk in dirty fossil fuel power plants as well as nuclear power. When the Electricity Feed-In Act was introduced in 1991, it could consequently be characterized by its sponsors in the country's two main parties as adhering to a long West German tradition of Ordoliberalism, or social market economics, within which the state acts to promote and maintain a healthy level of competition as a counterweight to capitalism's inherent tendencies toward monopoly.[427] Like PURPA in the United States, the Feed-In Act obliged grid companies to connect all renewable power plants to the grid. It also paid generators of renewable energy a guaranteed fixed rate for the electricity they generated through twenty-year power purchase agreements. Although the act helped stimulate the rapid growth of wind power in Germany, deployment of solar power was initially disappointingly low, a fact that reflected the low levels of remuneration for solar at the outset. Nevertheless, after the plan was revised under a coalition government led by the Green Party in 2000, a rapid increase in the deployment of modern renewables in general was sparked. The German law was so successful that the Feed-In Tariff scheme became the model used to spur renewables deployment across the European Union.[428] The scheme was remarkably effective: between 2004 and 2011, investment in clean energy in the EU rose by 600 percent.[429]

But the radicalism of Germany's *energiewende* lay not simply in the fact that renewable energy surged into the grid in unprecedented quantities, surprising and unsettling the country's big utilities and political establishment. As we saw in the last chapter, the rollout of renewable energy can take place in a manner that exacerbates existing economic and social inequalities, rather than establishing new forms of energy commoning. The revolutionary nature of the energy transition in Germany, by contrast, lay in *who* was generating renewable power. By 2012, nearly half the country's investments in new solar, wind power, and biomass (energy derived from waste produced by food crops and animal farming) came from German citizens and energy cooperatives; institutional investors like banks, insurance companies, and municipal energy suppliers constituted almost all of the other 50 percent—the country's big four private electric utilities (E.on, RWE, EnBW, and Vattenfall) were only responsible for 5 percent of renewable power.[430] The *energiewende*, in other words, really was an example of power flowing from and to the people. The history of extra-parliamentary struggles by German Autonomen groups against nuclear power and power dictatorship more broadly was a key prelude to this assertion of autonomous power.

The tiny village of Feldheim in the countryside outside Berlin has become a symbol of the empowering character of the German energy transition.[431] Like many other rural villages,

Feldheim had a lot of potential wind power. Individual farmers could have taken advantage of this by signing sweetheart deals with big renewable energy companies, reaping profits from wind power contracts while pissing off their neighbors—who would be constantly reminded of their exclusion from this windfall by the turbines that loomed over their village. Instead of adopting this divisive and individualistic strategy though, the residents of Feldheim decided in the mid-1990s to negotiate collectively with a startup renewable power company. Village residents decided together about the rules for fair distribution of the proceeds generated by the eventual field of forty-five giant wind turbines that were set up just outside the village. Their plan involves not simply economic benefits for all members of the village but also services such as a lighting system for the local football field. Feldheimers also worked with the developer to build a biogas plant that provides heat for houses in the village using a slurry of unused corn and pig manure, an arrangement that generated local jobs as well as benefiting the village's farmers.[432] The same year that they opened their biogas plant, the village of 150 residents decided to take control of the grid—a logical decision given that the village was producing all of its own energy. But the big utility that owned the grid in the region, a company now known as E.on, refused to sell or lease to the village. In response, the villagers pooled their money and asked the renewable power company they had been working with for nearly a decade to help them build their own smart grid. The process took two years and cost each of the town's 150 inhabitants just under $4,000, but now Feldheim has won complete energy autonomy and self-governance. Villagers pay 31 percent less for electricity than they did before their break with the power dictatorship. While not every town has the same felicitous combination of abundant winds and social cohesion with which Feldheim is blessed, the village has become an example of energy democracy for the rest of Germany and receives regular delegations from around the world.

Big energy companies like E.on were taken by surprise by the quick uptick in renewables following passage of Germany's renewables law of 2000. They did not wait long to begin pushing back politically. Germany's feed-in tariff scheme paid generators of renewable energy rates well above the retail price of electricity. The difference was made up for through government funding. Following the Great Recession of 2008, the big utilities and pundits on the Right began arguing that a government forced to impose austerity could no longer afford the scheme. Seeking to gin up populist sentiment against renewables, they also argued that feed-in tariffs are economically regressive—that they transfer economic resources from the vast majority of energy consumers toward investors, private renewables companies, and relatively wealthy homeowners and businesses that could afford—and had room to install—significant solar power.[433] They were not entirely incorrect on this score, and their arguments resonated to a certain extent with the public as tariff payments passed on to energy consumers raised residential electricity bills by more than 15 percent in 2012. By 2014, tariff payments to support renewables amounted to about a quarter of the average household electric bill in Germany.[434] Newly ascendant Far Right parties like the Alternative for Germany (AfD), learning from strategies deployed in the United States, began arguing that laws encouraging the growth of renewables are a form of "climate change hysteria" and "eco-dictatorship" that harms the average citizen.[435] This anti-renewables rhetoric has spread across Europe: in the spring of 2019, Finland's anti-immigrant Finns Party warned that progressive environmental policies would "take the sausage from the mouths of laborers," echoing Trump's ridiculous claim that the Green New Deal would rob Americans of their hamburgers.[436]

Stung by this reactionary populist rhetoric, European nations have moved to end feed-in tariff schemes and, in their place, embraced a system of competitive bidding in which generators of renewable power must compete against one another to win contracts from the state.[437] The goal of these competitive bidding laws is to cut the cost of renewable power but the effect has been to dramatically undermine the decentralized and democratic character of the energy transition since small renewables companies and community-based cooperatives generally lack the capital to submit artificially inexpensive bids. The result has been a shocking crash in levels of investment and deployment of renewables: as feed-in tariffs were replaced by competitive bidding schemes between 2014 and 2016 in Germany, investment in renewables fell by over 60 percent.[438] Across the EU, investments in renewables plummeted precipitously, dropping by more than half between 2011 and 2015. Despite the increasing cheapness of renewable power, that is, investment and deployment of clean power has not kept up as market incentives have replaced explicit state support. Although zombielike neoliberal rhetoric about a market-led energy transition lives on, with big banks predicting an investment boom as renewables become competitive with fossil fuels, Germany is currently not on track to meet its ambitious goals of generating 65 percent of its energy from clean power by 2030. In order to meet the climate goals it agreed to in Paris, Germany would have to commission roughly 1,700 wind turbines with a combined output of 5,000 MW to 6,000 MW each year, but according to Greenpeace Energy, only 600 MW were installed in the first three quarters of 2019.[439] Despite the phenomenal victories of energy democracy activists over the last several decades, Germany's celebrated *energiewende* was suffering a grave crisis as the second half of the decade began. People's power had made great strides but it was now evident that this success was only possible as a result of state support. The question of public power, it became clear, was unavoidable.

ENERGY DEMOCRACY IN BERLIN AND THE QUESTION OF STATE POWER

Berlin has been at the forefront of struggles for what might be termed common infrastructure for the last two decades. In 1999, Berlin's Senate, the city's governing body, signed a twenty-nine-year contract with the French transnational water corporation Vivendi and the German energy behemoth RWE that partially privatized the city's water authority.[440] According to this agreement, the city-state of Berlin held 50.1 percent of the shares of the newly created water utility while Vivendi and RWE held 49.9 percent. The terms of this so-called public-private partnership (PPP) guaranteed high rates of return to private investors, leaving the citizens of Berlin on the hook if the new water utility failed to generate expected big profits. Reacting to this potentially disastrous socialization of risk and privatization of profit, social movements in Berlin launched a public referendum in 2011 that forced the city to re-municipalize its water supply. A prominent image for the campaign drew on the famous symbol of the country's antinuclear movement to satirize the financial sharks seeking to profit from the partial privatization of the city's water supply. The re-municipalization of Berlin's water was part of a much broader trend to seize back public control of vital social infrastructure. A report from the Transnational Institute on this global wave of re-municipalization reveals that 347 out of the total 835 examples covered took place in Germany, and that 284 of these German cases were related to energy.[441] Having reversed the privatization of their city water supply, Berlin's citizens were determined to take back public control of their energy.

Berlin sold off its energy utility in 1997 following the liberalization of the German electricity market in the 1990s. At the time, the city's economy was in decline and public budgets were squeezed, but, in addition, it had become neoliberal dogma that government management of services was inefficient and monopolistic. Backed by powerful financial institutions like the International Monetary Fund and the World Bank, the so-called Washington Consensus held that privatization and deregulation of public infrastructure services would force them to economize and to be more responsive to consumer needs. In the developing world, this neoliberal doctrine was enforced by strict conditions that tied future loans to privatization of public infrastructure and austerity, while in the EU neoliberal dogma led to a continent-wide privatization of electricity suppliers in the mid-1990s. The idea was that government support in the form of feed-in tariffs would push nimble investors toward renewable energy, generating a synergistic relationship between the free market and energy transition. It didn't work out that way. Berlin's public utility was bought by the Swedish state-owned conglomerate Vattenfall, a company with large holdings in nuclear power and in Germany's highly polluting lignite coal mines. After nearly two decades in control of the city's grid, the Vattenfall subsidiary Stromnetz Berlin GmBH had a dismal environmental record: Berlin ranked last in terms of renewable energy development in comparison to all fifteen other federal states in Germany, two others of which are city-states like Berlin.[442] For environmental activists in Berlin, it was clear that Vattenfall's large investments in nuclear power and fossil fuels were "stranded assets" that prevented it from making the switch to clean energy. It also didn't help that for-profit power companies like Vattenfall have every incentive to encourage people to consume rather than to conserve energy—particularly given Germany's stated goal of cutting energy consumption by 20 percent from 2008 levels by 2020.

Activists moved to take Berlin's grid back into public hands. In the summer of 2011, an alliance of fifty-six local civil society groups founded the *Berliner Energietisch*, or Berlin Energy Roundtable.[443] With the concession contracts of the city's energy and gas supply due to expire in 2013 and 2014, the groups in the *Energietisch* sensed a window of opportunity for re-municipalization. After a three-month-long series of meetings and

discussions, participants in the *Energietisch* finalized a draft
law for a public referendum in January 2012. The law, called
"New Energy in Berlin: Democratic, Ecological, Social," contained two basic demands: that Berlin should put forward a bid
to take back its electric distribution network from Vattenfall; and
that the new municipal utility company—to be called *Berliner
Stadtwerke* in German—should be publicly owned by the local
state on a nonprofit basis. The latter elements of the proposal
were particularly exemplary inasmuch as they incorporated
all of the fundamental ingredients of energy democracy iden-
tified by a subsequent study of European movements by the
Rosa Luxemburg Foundation.[444] Animated by the terrible environmental record of Vattenfall's Berlin subsidiary, members of
the *Energietisch* specified that the new municipal energy utility
should sell 100 percent clean energy, and that it should invest
whatever profits it made in new renewable generating capacity
for Berlin. Rather than being siphoned out of the city, and, in the
case of Vattenfall, being used to support destructive expansion
of dirty fossil fuel infrastructure, profits would remain in the
local economy and would help support a virtuous cycle of conversion to clean power. In addition, the *Energietisch* specified that
Stadtwerke would provide affordable energy to the citizens of
Berlin and would tackle energy poverty by charging lower rates
to economically disadvantaged residents of the city. The corporate logic of profit maximization, it had become clear, militated
against both the goal of switching to renewable energy in order to
avert climate emergency *and* against meeting basic social needs,
particularly those of the more vulnerable residents of the city.

In order to generate higher profits, private utilities must charge customers higher rates, a dynamic that had begun inflicting significant economic pain on segments of the German population, which in turn became vulnerable to the antienvironmental rhetoric of the Far Right. Re-municipalization offered an antidote to this dynamic by simultaneously cutting energy consumption, speeding up the transition to renewables, and addressing energy poverty. For energy democracy campaigners in Berlin such as Michael Efler, energy should be seen as a human right not a commodity, and access should be granted to all regardless of their means.[445]

The *Energietisch* proposal constituted a momentous upending of the logic that had led to privatization of Berlin's energy utility. But there were nevertheless concerns among some residents of Berlin that the outcome of re-municipalization would not necessarily be as beneficial as activists envisaged since the new utility would be managed by a relatively insular and bureaucratic state elite. As a result, a group of Berliners launched a parallel initiative to the Energy Roundtable, forming a citizen-owned energy cooperative called *BürgerEnergie Berlin*, or Berlin Citizen's Energy. According to *BürgerEnergie* founder Luise Neumann-Cosel, the goal of the cooperative was to put the grid back directly into the hands of citizens rather than under the authority of the local state.[446] Neumann-Cosel and her comrades in the *BürgerEnergie* initiative worried that decisions made by the city were shaped by a powerful senator for finance whose outlook they believed was indistinguishable from that of a corporate CEO.[447] Like the new public utility proposed by the *Energietisch*,

BürgerEnergie planned to reinvest profits in social and environmental causes, but some of those profits would be redistributed to shareholders, a form of privatization of the social surplus that irked the more Leftist members of the *Energietisch*.[448] At the same time, organizers of *BürgerEnergie* recognized that not everyone could afford to (or would want to) buy shares in order to become a member of the cooperative. There were, in other words, unmistakable limits to the radical participatory ethos of the energy cooperative. Members of *BürgerEnergie* consequently pushed for only partial ownership of the grid: they always wanted a majority of the utility to be owned by the city in order to ensure that the strongest vote in all governance matters should be with people openly elected to represent the interests of *all* of the citizens of Berlin.[449]

For its part, the *Energietisch* was also intent on developing new and more democratic mechanisms of governance for the new utility. A prominent slogan of the campaign in fact was *Power to the People*.[450] To ensure genuine citizen input, the Energy Roundtable specified that *Berliner Stadtwerke* would be governed by participatory democratic mechanisms that included elected citizens and workers on the managing board, advisory open neighborhood assemblies, and the ability to force the board to discuss issues in response to public petitions.[451] Instead of delegating their votes to the city's governing body, the Senate, citizens would constitute six out of the fifteen board members making decisions about the running of *Stadtwerke*, an arrangement that it was hoped would ensure the prioritization of affordable electricity as well as precluding the reprivatization of the new public utility. The proposals from the *Energietisch* were essentially experimenting with new forms of public ownership in which direct democracy played a crucial role.

As the scholar and energy democracy activist James Angel explains in his fascinating account of the *Energietisch*, for some Left activists involved in the initiative, the state was not to be discounted as a "mere instrument of capital." While it is heavily influenced by the demands of private capital, for these activists the state "remains a fruitful terrain of struggle." Angel quotes one activist from the Interventionist Left formation that argues that the state is "a condensation of power relations [. . .] something which is designed to negotiate the hegemony around the capitalist class" while "reflect[ing] the struggles in society." These views were certainly shaped by the often-fruitful relationship between radical extra-parliamentary groups such as the Autonomen and insurgent groups working within the existing political system such as elements of the Green Party. In addition, as Angel notes, *Energietisch* activists were also influenced by the tradition of theorizing about state power that extends from the Italian revolutionary Antonio Gramsci to the France-based theorist Nicos Poulantzas. In his seminal book on state power, Poulantzas argued that the tradition of theorizing state power that descended from Lenin and the Third International, as well as the political positions that flowed from these theories, had reached an impasse.[452] According to Poulantzas, this tradition had come to see the bourgeois state as a monolithic (and necessarily oppressive) bloc. Lenin's bitter experiences during the Russian Revolution led him to conclude that bourgeois representative democracy was a façade that would never make meaningful concessions to the forms of direct, rank-and-file democracy developed in the workers councils or Soviets. Lenin and his followers concluded that the struggles of the popular masses therefore could never pass through the state but necessarily took the form of the creation of a dual or alternative power structure. Taking power involved destroying the machinery of the bourgeois state and replacing it with workers councils. For Poulantzas, this permanent skepticism about the state precluded the possibility of mass intervention in existing politics, generating a jaundiced theoretical outlook that he argued slid remorselessly into distrust of popular movements themselves.[453] It was, after all, the Communist Party rather than the Soviets that ultimately took power once the bourgeois state had been smashed. Interestingly, Poulantzas argued that there was a fundamental correspondence between this Stalinist statism and post-1945 European Social Democracy, which he saw as also being animated by a fundamental distrust of direct democracy and popular initiative. Social Democrats, he suggested, believed that "occupation of the state involves replacing the top leaders with an enlightened left elite," who would "bring socialism to the masses from above."[454]

While Poulantzas was writing in the 1970s and articulating antiauthoritarian sentiments that characterized many other thinkers of his era, his critique of Left theorization of the state as a monolith resonates powerfully today. After all, the tradition from which the Autonomen emerge tends to rigorously eschew the taking of state power in favor of the creation of radical autonomous movements and spaces that are intended to pre-figure broader social transformation. Moreover, the Autonomen movements in Germany are part of a much broader, global trend. From Argentina in 2001 to the Zapatistas and Occupy Wall Street, Left social movements of the last few decades have tried, in John Holloway's phrase, "to change the world without taking power."[455] These movements tend to believe that social change is terminally deferred by demanding reforms of the state or by try-ing to institute reforms through the incremental taking of state power.[456] As we have seen, however, it was precisely through the creation of public incentive programs that hundreds of rad-ically democratic energy cooperatives were able to bloom during Germany's *energiewende.* When these state-based incentives were withdrawn, the torrent of grassroots-based clean power was reduced to a trickle. The German *energiewende* experience thus offers empirical confirmation of Poulantzas's theoretical points about state power. For Poulantzas, the key question was: "How is it possible radically to transform the state in such a man-ner that the extension and deepening of political freedoms and the institution of representative democracy (which were also a conquest of the popular masses) are combined with the unfurl-ing of forms of direct democracy and mushrooming of self-man-agement bodies?"[457] For Poulantzas, in other words, the state is a heterogeneous formation: one wing is undeniably repressive, primed to crush movements that challenge the power of elites; there are, however, other sectors of the state, including regula-tory agencies and redistributive organs, that are a product of past struggles for social, economic, and environmental justice. While these latter sectors are often subject to the dictates of the powerful, they are also inclined to be responsive to the demands of mass social movements. Poulantzas enjoins movements to cultivate powerful autonomous political forms (direct democracy and self-management), but then and in tandem to use the momentum developed through these autonomous institutions to exert power on susceptible bodies within the state. Getting this combination right is obviously a delicate dialectical dance: there is no simple formula to avoid the co-optation of autonomous movements by elite-controlled state organs, or, conversely, to prevent the insistence on self-management from freezing out efforts to take part in what New Left activists called "the long march through the institutions." But, in the best instances, this dialectical dance can prove a formidable combination that can fertilize what Poulantzas calls "a mushrooming of democratic organs at the base"—a kind of proliferation of practices of communing— while simultaneously shifting the relation of forces within the state in a progressive direction.

THE ROAD TO VICTORY—AND WHAT TO DO AFTERWARD

So what happened to the *Energietisch* proposal for remunicipalization of Berlin's energy utility? Predictably, Vattenfall fought it with all sorts of dirty tricks, and many local politicians in Berlin supported the company in its campaign to thwart the *Energietisch*. By the middle of 2013 *Energietisch* activists had gathered over two hundred thousand signatures in favor of a popular referendum on remunicipalization. With just four days to go before the referendum, and in the face of vociferous protest, the Senate of Berlin shifted the date of the referendum vote away from the day of the general election in a move calculated to diminish voter turnout. Notwithstanding this maneuver, six hundred thousand people voted in favor of remunicipalization of Berlin's energy supply in November 2013, an 83 percent majority. Despite this massive support for the *Energietisch* proposals, the referendum fell twenty-one thousand votes short of the required voter quorum and so did not go into effect. This depressed vote can be explained partially by another dirty trick deployed by the Senate: ten days before the referendum vote, Berlin's governing political coalition of Social Democrats and Christian Democrats announced the creation of a mini-municipal electric utility. This move incensed many *Energietisch* activists, who rightly saw it as a transparently cynical move by members of the conservative governing coalition with a longstanding record of opposition to remunicipalization to diffuse support for the *Energietisch* proposals. The new mini-utility, which was named the *Stadtwerke*, operated with severe constraints since it was prohibited from trading energy other than that which is generated through its own investment, which the Berlin Senate has insured is anemic.[458] Worse still, the new utility was shorn of the elements of social justice and radical democracy that had been central to the *Energietisch* campaign—elements calculated to make the utility a pioneering model of a new form of public power.

This defeat was undeniably stinging, and yet all was not lost. For Michael Efler, who at the time of the referendum was working for the nonprofit organization *Mehr Demokratie* (More Democracy) but who was subsequently elected to the Berlin regional parliament as a member of the *Die Linke* (Left Party), the years since the 2013 referendum have seen significant successes for the movement for public power. Once in parliament, Efler managed to work in coalitions to implement what he sees as 80 percent of the original *Energietisch* demands, including their positions on clean energy and the decommodification of energy. While the Social Democratic party has stonewalled the elements of participatory governance demanded by the *Energietisch,* giving support to Poulantzas's assessment of the elitist and technocratic character of the organization, Efler told me that he and his comrades were continuing to advance measures for better parliamentary control of the *Stadtwerke.* In addition, the *Energietisch* campaign ultimately strengthened struggles for energy democracy in the city. For Green Party activist Oliver Powalla, the *Enquete,* or official inquiry, that was launched following the referendum was an important opportunity to chart progressive directions in Berlin's energy politics.[459] The result was a series of far-reaching proposals for quitting coal, dealing with energy poverty, and managing the Berlin energy grid that the conservatives in government were forced to address and implement. Moreover, the fight against Vattenfall and the political establishment opened up other campaigns in the surrounding region. Powalla is now part of a campaign to make the costs of coal production in the state of Brandenberg, which surrounds Berlin and is Germany's second-biggest producer of coal, visible to the citizens of Berlin.[460] The campaign involves making Berliners aware of the polluting impact of coal pollution on the Spree River, which is the city's drinking water source. In the process, Powalla says the movement is forging valuable links between city-dwellers and rural folk, reviving precisely the kinds of solidarities that fueled the mass campaigns against nuclear power in the 1970s and 1980s.

On March 5, 2019, the Berlin Senate announced that the state-owned company *Berlin Energie* would be awarded the contract for the power grid license for the city. This decision was the culmination of six years of struggle by energy democracy activists in the city to take the grid back into the hands of the public. Although the electricity network license had expired in 2014, Vattenfall successfully delayed the switch to public power for half a decade using tactics of legal obstructionism. In a press communiqué released the same day as the Senate's decision, the *Energietisch* stated, "This is a very good day for the 600,000 Berliners who voted for a power grid in Berlin in 2013 in the referendum. The profits from network operation are finally back in town. The central goal of the municipal grid operator must now be to advance the switch to 100 percent decentrally generated renewable energies and to implement the conversion and expansion of the power grid accordingly. We expect that the provisions on transparency and participation in the draft referendum will be implemented swiftly."[61] This was a huge victory for the long campaign waged by the many activists in the *Energietisch*, but as their press release suggests, vigilance is still required to make sure the new public utility implements the campaign's goals of energy transition and democratic participation adequately.

Moreover, the struggle for public power is opening up other fronts in unexpected zones. For instance, as Oliver Powalla explained, one of the new areas of popular struggle linked to

energy relates to housing.[462] After all, Germany's dirty coal
reserves don't just generate electricity: they also produce heat for
homes and public buildings. As the country moves to phase out
coal by 2030, a battle is intensifying over which energy sources
will heat the nation's buildings. This is a huge issue since residential heating accounts for 40 percent of energy-derived carbon
dioxide emissions. Once again, Vattenfall is in the middle of this
conflict since it owns a massive centralized heating system that
sends heat to 1.2 million housing units in Berlin. The good news is
that district heating systems are vastly more efficient than ones
based in individual homes. Such systems are typically fired by
so-called Combined Heat and Power (CHP) plants, which generate electricity as well as heat in a process far more efficient than
pure power plants—which typically lose up to two-thirds of the
energy they use as "waste" heat. But will Vattenfall switch its district heating system from coal to natural gas? Although the latter
is undeniably less polluting than coal, it is still a fossil fuel. The
Energietisch movements are becoming involved in this debate
over the future shape of heat in Berlin. As was true of the struggle
over the grid, this building battle ties struggles over energy transition to issues of economic and social justice. Energy democracy
activists in Berlin want to make sure that district heating systems
in the city are powered by renewable energy, from excess wind
power to geothermal, biogas, and other alternative sources. But
they also want to make buildings more efficient so that it takes
less power to heat them. This ecological measure would also
save money, but how will such efficiency measures be paid for?
If increased rents pay for efficiency measures, an anti-ecological

populist backlash may be sparked. But if done right, in a manner that takes social inequality into consideration, insulation of buildings and other efficiency measures could make it cheaper to heat buildings and thereby help to bring down the spiraling cost of housing in Berlin. Vattenfall does not necessarily share any of these objectives since it wants to make as much profit from its district heating system as possible. Energy democracy activists in the city of Hamburg, who in November 2018 bought back their district heating system from Vattenfall, have perhaps already forged the path forward for Berlin. This remunicipalization was driven by plans to power the city's district heating entirely with renewables.[463]

LESSONS FOR PUBLIC POWER

What are the lessons of the *Energietisch*'s struggles for other movements fighting for energy democracy and public power? First and most crucially: David can beat Goliath. The campaign waged by Berlin's *Energietisch* demonstrates that an alliance of social movements *can* prevail against a politically connected and deep-pocketed transnational corporate behemoth. In order to do so, however, the movement had to hammer on three interlinked points. First of all, the private utility had stranded assets in fossil fuels that prevented it from making the transition to renewable power with the speed and on the scale necessary to avert climate chaos. Second, a switch to public power would lower people's utility bills and would allow considerations of social equity to figure in decisions about power—no more disconnecting people from power because they can't pay the utility's inflated rates! Lastly, public power needn't and should not resemble the unresponsive technocracy of the bad old days. New forms of public power are possible in which popular movements and the citizenry in general can exercise greater direct democratic control over the utility in particular and over power in general.

The activists who gathered in a public school in Queens on that muggy summer night in 2019 began their campaign for public power in New York City from precisely these interlinked points. Con Ed rates are already too high, they claimed, and it has repeatedly failed the people of New York in their time of most need, like during the heat wave that summer. The big for-profit utility model is an impediment rather than a catalyst for energy transition. And we, the people of New York, are sick of the corruption and revolving door deals between Con Ed, state regulators, and political elites in the city and state. The struggle for energy democracy in New York is going to be epic. Con Ed is every bit as much of a goliath as Vattenfall. And there is far less of a consensus in the city and around the country regarding the necessity of energy transition. But the tide of public opinion can often change remarkably quickly. The struggle for public power in New York has now been folded into the movement for a countrywide Green New Deal, and activists are mobilizing behind a sweeping set of demands for change.[464] Energy transition is undeniably the core of this plan for the ecological and social reconstruction of the United States and the world beyond.

Today the question of public power is unavoidable. We need to scale up the transition to clean energy massively. The only way to do this, it is now clear, is through a thoroughgoing decommodification of energy through public procurement. As Sean Sweeney and John Treat of Trade Unions for Energy Democracy point out, making the transition to renewables in the short time we now have left requires two key ingredients: long-term planning and abundant and steady financing.[465] Privatization and liberalization of energy utilities militated against both of these ingredients by siphoning money away from the funding and deployment of renewable infrastructure to private pockets, and by creating a scrambled and competitive institutional landscape dominated by large "power dictatorships" whose "sunk assets" pushed them to put the brakes on the energy transition. By contrast, public power can generate funding that is used to build out the renewable grid and complementary social infrastructures that reduce energy consumption such as public transport and zero-energy housing. And the radical new forms of public power proposed by *Energietisch* activists can facilitate long-term planning with genuine mass public support.

Time is short. We must be ambitious in our goals and radical in our demands. All power to the people!

CONCLUSION:
INFINITE ENERGY?

T. X. Watson's short story "The Boston Hearth Project" narrates the story of a raid by a team of hackers on a swank building the city of Boston is set to open in a near future of dystopian climate breakdown. Watson's story, which is part of the solarpunk anthology *Sunvault*, is narrated by a teenage computer gaming champion named Andie Freeman, who decides to use his skills during his gap year before college to good activist ends. Early in the story, he identifies the two key developments that animate his activist work:

> 1. The rate of winter deaths of homeless people in Boston has been increasing every year since climate change has made weather patterns more and more erratic [. . .] 2. The city of Boston built, and was about to open, a new living building: the Hale Center. It was a big fifteen-story Art Deco Revival temple with a custom-engineered closed ecosystem. It was basically a first-class hotel, set up so that business people and politicians wouldn't have to go around interacting with the actual city.[466]

As Freeman explains, the building is autonomous and impermeable to the external environment, a self-sufficient solar-powered refuge for elites intent on escaping the environmental and social collapse generated by a rapacious capitalist system: "The air in the Hale Center is different than the city outside it. It smells like being outdoors, somewhere other than Boston. There's no smell of cars and mud and construction. There's no recycled-air-and-cleaning-products smell. There's a breeze, and it carries the varied and pleasant scents of the building's miniature biomes."[467] Freeman uses his gaming skills to guide a teammate named Juniper Berg, a supremely physically skilled "urban explorer" and *traceuse* (parkour aficionado), as she breaks into the Hale Center, hacks its computer security and control systems, and opens the building to a group of activists and homeless people. Once they have taken it over, the living building's sophisticated systems become "our weapon as well as our hostage," allowing the urban insurgents to hold the building against a police siege for forty-nine days. The insurgents' social media team drums up massive public sympathy, and eventually the city is forced to convert the building into the Boston Hearth Homeless Shelter. An elite lifeboat is thereby turned into a sanctuary for the city's most marginalized denizens. As Freeman explains near the end of the story, "having a building that's designed to turtle against the outside world makes an immense difference, and we're saving a lot of lives."[468]

Watson's short story imagines the role activists can play in transforming elite infrastructures of climate adaptation, rendering them part of an egalitarian urbanism for the age of environmental breakdown. In this regard it is symptomatic of the emerging subgenre of solarpunk, which, according to one critic, is just as much a growing community and an expression of radical political desires as it is a literary genre.[469] Solarpunk stories are thus "images of post-oil futures," as well as stories of "community, of kin, of friendship and care, and of strife overcome by communal striving."[470] They share much with other forms of speculative fiction, which conjures up imagined worlds won through political activism. In the case of solarpunk, however, imagined futures pivot on questions of just energy transition. Given the oppressive legacy of fossil capitalist infrastructure, solarpunk inevitably stirs up discomforting questions about what the physical and social infrastructures of future societies organized around energy democracy will look like. What, in other words, will be the materiality of solar energy, and how will a society organized around solar power—a condition that has been called "solarity"—be organized? Solarpunk as a genre returns us to many of the questions posed by Gifford Pinchot and his comrades as they wrote about and strategized for the advent of "giant power," or universal electrification. Now as then, it is clear that the control of energy confers power and catalyzes social transformation. The narratives spun by solarpunk writers are efforts at designing alternative futures, openings to possible egalitarian arrangements of energy, infrastructure, and society.

Yet this celebration of energy transition generates what cultural critic Rhys Williams calls "amnesia of solar infrastructure."[471] For Williams, "the sheer intimacy and ubiquity promised by the affordance of aestheticized solar technologies like

these causes them to become functionally invisible, whether
through transparency or through an aestheticisation that con-
ceals their technical purpose."[472] Unlike leaves, after all, solar
panels do not grown on trees: they must be manufactured, using
chemicals that are often highly toxic and in conditions that do
not escape the conditions of labor exploitation and degradation
that characterize the era of fossil capitalism. To conceal these
conditions of exploitation, designers of solar cells are intent on
improving the aesthetic qualities of solar power, making transparent cells that render solar infrastructure invisible. But, like all
other energy infrastructures, solar power can only remain invisible to certain (relatively affluent) people, and only for a limited
period of time. As Williams puts it, "Solar technology provides
energy autonomy, but only for as long as the product works, only
for the lifespan of a solar cell, before the question of production
arises again."[473] Eventually, the question of controlling the means
of production returns.

T. X. Watson's "The Boston Hearth Project" illustrates the
bind in which much solarpunk finds itself. In breaking into and
taking over the Hale Center, Andie Freeman and his comrades
seize the means of solar production. The building after all is
self-sustaining and can exist apparently autonomously not just
of the electrical grid but also of the surrounding environment
in its entirety. Their physical and digital infiltration of the Hale
Center, and subsequent repurposing of the building as a homeless shelter, is very much in line with the long tradition of cyber-
punk, in which subversive and marginal hackers are able to take
over the means of digital communication and use this control to

overturn a repressive social order. It is also worth noting that this insurgency is very much in line with the transformation Hardt and Negri imagine the cognitariat engaging in, as they use capitalism's means of digital production to establish new forms of commoning. But do Freeman and his comrades really control the means for producing solar power? What will happen when the solar panels on the Boston Hearth Project break down? Will the hackers and homeless coalition figure out ways to build their own solar panels to replace the ones that stop working? If so, where will they acquire the silicon, copper, and other resources that make up solar modules, not to mention the steel or aluminum that are used as frames for the modules? These thorny questions constitute what might be called the political unconscious of solarpunk, its tendency to ignore questions not simply of control over the means of production but also of resource flows and the neocolonial power relations between peripheral sites of extraction and the imperial centers of capital that shape the world system today. This is not to suggest that energy sovereignty is a negligible goal. The community-owned streetlights put up by Soulardarity in Detroit and the solar commons established in NYC and Berlin are genuine victories of popular struggles for energy democracy, but the political unconscious of solarpunk reminds us that such victories are ultimately reliant on the very relations of production that they are striving to overthrow.

An energy transition that maintains existing forms of capitalist production and infrastructure will be nothing short of devastating for the planet and vulnerable frontline communities. In a little-remarked upon report from 2017, the World Bank

modeled the projected increase in resource extraction required
to build enough solar and wind utilities to produce an annual
output of seven terrawatts of electricity by 2050—enough to
power only about half the global economy.[474] As the report notes,
"the technologies assumed to populate the clean energy shift—
wind, solar, hydrogen, and electricity systems—are in fact significantly MORE material intensive in their energy composition
than current traditional fossil-fuel-based energy supply systems."[475] In his commentary on the report, anthropologist Jason
Hickel observes, "in some cases, the transition to renewables will
require a massive increase over existing levels of extraction."[476]
Even more startling, the report does not include any figures for
other sectors that need to be electrified such as transportation
and the heating and cooling of buildings. Granted, e-cars running solely on electricity produced from modern renewables will
have a lower carbon footprint over their lifetime than internal
combustion vehicles powered by gasoline; but their production
still requires loads of energy and resources, including some that
are predominantly located in the Global South.[477] And replacing
the world's projected fleet of two billion vehicles with e-cars will
require an explosion of additional mining of key minerals, with
extraction of copper needing to more than double and cobalt to
quadruple by 2050. As Jason Hickel observes, it's not that we're
going to run out of these minerals. Instead, the projected massive increases in mining will intensify an already grave global
crisis caused by extractivism, which is the single largest driver of
global deforestation, ecosystem collapse and biodiversity loss, as
well as the catalyst for systemic human rights abuses. The fantasy

of a limitless solar economy is thus an intensification of the capitalist fairy tale of limitless freedom so wickedly embodied in car commercials, with their images of men driving through sublime natural landscapes with reckless abandon.

The big question raised by most accounts of energy transition, one that is all too seldom broached, is whether we actually need to maintain the current high energy, high consumption capitalist economy. As *Land Institute* scientist Stan Cox points out, none of the dominant scenarios for going fossil-free include discussions of strict regulations on the amount of energy consumed in production and consumption in rich nations like the United States.[478] This is as true for campaigning groups like 350 .org and Fossil Free as it is for influential institutions like the IPCC.[479] But why should we hope for a fossil-free society that consumes just as much power and looks just like the present fossil capitalist order? The vertiginous inequalities that characterize contemporary global capitalism ensure that energy consumption is totally lopsided within and between nations, with one in five people lacking access to modern electricity around the world.[480] Meanwhile, the global rich are quite literally consuming the future through their unbridled appetite for energy: fully half of global carbon emissions come from the wealthiest 7 percent of society.[481] In a nominally wealthy global city like New York, over a fifth of city residents live below the federal poverty line, nearly eighty thousand people are homeless, and 27 percent of residents skip meals because they don't have enough money to buy food. All of this while tycoons like former mayor Michael Bloomberg earn billions of dollars every year and hedge fund managers spend hundreds of millions of dollars on homes in which they will rarely live.[482] Meanwhile, fossil capitalist society ensures that we work ourselves silly at what the anthropologist David Graeber poetically calls bullshit jobs, most of which simply help prop up markets so that we will continue to overconsume useless, soul-eating, and planet-destroying stuff.[483]

Most calls for a shift to 100 percent renewable energy focus on either the market dynamics or the technical aspects of the shift while ignoring the underlying social structures and belief systems that lead to high energy consumption in the United States and other wealthy nations. They thereby unconsciously help to maintain the status quo of societies that were designed to maximize unsustainably high consumption. In the United States, for example, government worked hand in hand with automakers and real estate developers in the post-1945 period to develop the sprawling automobile-highway-suburban complex that has come to seem normal to most Americans—and that is being exported with disastrous environmental consequences around the planet.[484] There are currently over a billion cars in the world, a number projected to double by 2040.[485] Since car ownership is responsible for about a quarter of the average American's personal carbon footprint, which, at over sixteen metric tons annually, is one of the largest in the world, the expansion of the automobile complex to countries like India and China has lethal implications for the planet. Would banning the combustion engine and providing free electric vehicles instead solve this problem? Not at all, since the manufacturing of a car—an immensely complex enterprise that uses mountains of resources (lithium, rubber, plastic, metal, paint, etc.)—creates as much carbon pollution as driving it does.[486] Building an average car generates about seventeen metric tons of carbon dioxide, and luxury SUVs produce double that.[487] Simply replacing current gasoline-powered cars with electric vehicles would eliminate none of these emissions. In other words, even if combustion engines were banned today, and the two billion people that reports predict will own cars by 2040 bought electric vehicles, the world would vastly overshoot its carbon budget, tipping the planet into climate catastrophe.[488] Seeking to maintain the auto-highway-suburb complex is thus profoundly irrational on an environmental level, and this is without even considering the high cost of auto ownership and the 1.25 million people killed and 50 million injured in auto accidents around the world each year.[489] In place of such death machines, we need to fight for the highest quality forms of public transportation, including subway and tram systems in cities and high-speed intercity rail service.

Once we become aware of the immense amount of carbon produced by the construction of the average automobile, the question of industrial production in general rears its head. All too often, the climate crisis is blamed on some undifferentiated "humanity" (as in the popular academic notion of the Anthropocene, the age when "people" began decisively controlling the planet's geophysical systems). In the United States, this unfocused approach can all too often devolve into a browbeating lifestyle politics, in which individual consumer decisions rather than the capitalist system as a whole are blamed for the climate crisis. And yet industrial production consumes more of the world's energy than transportation, residential, and commercial sectors combined.[490] These statistics underline that the core problem the climate justice movement confronts is *controlling the means of production*. That means control of energy production specifically, but control of social production much more broadly as well. Public ownership of energy infrastructure in order to make a transition of the scale and scope necessary in the limited time left to us is absolutely necessary, but the question of public ownership of energy must be linked to broader questions about ending the current system of irrationally excessive expenditures, the cruel inequalities they generate, and the hollow ideologies of private affluence amidst public squalor that underpin them while switching over to renewable energy. Surely it would be better to transform society so that the bloated, conspicuous consumption of the wealthy is reigned in and basic energy is put within reach of all than to fight for a 100 percent renewable energy infrastructure that maintains this unjust and loveless system. Cuts in overconsumption of energy will certainly be necessary if we wish to substantially improve the living standards of the three billion people presently living in absolute poverty in the Global South.[491] Fossil capitalist culture has always been predicated on vertiginous growth, and the temptation today is to assume that we need more of this acceleration, that we need to hasten the transition to renewable energy in order to keep consuming more stuff. While this may be true on one level since we really do need to speed up production of modern renewables, the rhetoric of acceleration precludes us from focusing on how we might slow down other sectors of hyper-capitalist society. Do we really need all of the energy-hogging, resource-consuming, and polluting gadgets and baubles that consumer capitalism currently disgorges in such vast quantities?[492] And should we not question a capitalist system whose fundamental premise is inexorable compound growth or crisis?

There are presently just fewer than one billion people who lack access to electricity around the world. Although significant gains have been made in tackling energy poverty by nations such as India in recent years, the majority of the population in a vast region such as sub-Saharan Africa still lacks modern power.[493] In addition, 2.7 billion people around the world do not have access to clean cooking facilities, relying instead on biomass, coal, or kerosene for cooking, with serious health implications for women and children in particular. While simply hooking homes up to power does not necessarily translate into the empowerment of women since men often still exercise control over how that energy is used, electrification can be a vital part of a broader set of transformations that give women access to life-improving transformations, including education and economic empowerment.[494] The race for access to energy is on around the world, but that power is not always renewable. For example, China's colossal overseas development plan—known as the Belt and Road Initiative—perhaps the most ambitious infrastructure project in world history, has led to the establishment of at least 240 coal-fired power plant projects in twenty-five countries around the world.[495] Even when it is not actually building these dirty power plants, China is also financing about half of the proposed new coal capacity in countries like Egypt, Tanzania, and Zambia. The planet simply cannot afford to follow the same high-emissions pathway pursued by Western capitalist industrial powers over the last two centuries and by the Chinese more recently. This dirty energy threatens to lock out clean power while saddling poor countries with massive debt and increasingly destructive fossil fuel "stranded assets." Another road to coping with energy poverty is possible and necessary to avert planetary ecocide. The only way that this road is likely to seem credible to developing countries, however, is if the wealthy nations of the Global North not only adopt clean power themselves but also make the transition to renewables part of a much broader shift toward economies oriented around sustainable levels of consumption in general, and public control of power that generates jobs for the masses of people in the Global South who struggle to survive in various forms of precarious work.

To advocate a program of emergency contraction in wealthy nations is not to hope for the return of the horse and buggy, of locally woven carrot pants and pathogen encrusted compost toilets, but rather to challenge the idea of a future in which renewable power perpetuates an uber-capitalist world based on next-day shipping, global commodity flows, hyper-competition and exploitation, and inexorable, planet-destroying growth.[496] To transition successfully to 100% renewable power we also need to imagine an alternative, post-capitalist society, one focused not on continual accumulation simply for the sake of shareholders. Such a society would be grounded in popular planning focused on the genuine needs that go glaringly unmet today, including good jobs for all, a massive reduction of work time, universal affordable housing, functioning basic infrastructure, universal health care, and guaranteed access to basic energy needs for all. To make such a program of emergency contraction a political possibility, we need to organize a movement around demands not simply to go fossil-free but for a just transition and better, more fulfilling forms of living. The scale of the transformation we must make is daunting. But once we give up the empty hope that the free market system will save us and that we must maintain current feckless forms of "progress" and growth, we gain a sense of collective possibility that allows us to forge the solidarities that will redeem our current savage capitalist system. We have nothing to lose but the all-too-immanent prospect of extinction. In place of this chilling prospect, we must fight for a new world grounded in communal luxury and human solidarity.

BIBLIOGRAPHY

Abramsky, Kolya, ed. *Sparking a Worldwide Energy Revolution: Social Struggles in the Transition to a Post-Petrol World.* Oakland, CA: AK Press, 2010.

Abramsky, Kolya. "Introduction: Racing to 'Save' the Economy and the Planet: Capitalist or Post-Capitalist Transition to a Post-Petrol World?" In *Sparking a Worldwide Energy Revolution: Social Struggles in the Transition to a Post-Petrol World,* edited by Kolya Abramsky. Oakland, CA: AK Press, 2010.

Aden, Nate. "The Roads to Decoupling: 21 Countries Are Reducing Carbon Emissions While Growing GDP." *World Resources Institute,* April 5, 2016.

Adler, Ben. "How Are Environmentalists Reacting to Obama's Clean Power Plan?" *Grist,* August 3, 2015.

Alperovitz, Gar, and Johanna Bozuwa. "Electric Companies Won't Go Green Unless the Public Takes Control." *In These Times,* April 22, 2019.

Alperovitz, Gar, Joe Guinan, and Thomas M. Hanna. "The Policy Weapon Climate Activists Need." *The Nation,* April 26, 2017.

Amelang, Soren. "Battered Utilities Take on Start Ups in Innovation Race." *Clean Energy Wire,* May 16, 2017.

Anderson, Benedict. *Imagined Communities: Reflections on the Origins and Spread of Nationalism.* New York: Verso, 1998.

Andoni, Merlinda, et al. "Blockchain Technology in the Energy Sector: A Systematic Review of Challenges and Opportunities." *Renewable and Sustainable Energy Reviews* 100, no. 1 (2019): 143–74.

Angel, James. "Towards an Energy Politics In-Against-and-Beyond the State: Berlin's Struggle for Energy Democracy." *Antipode* 49, no. 3 (2017): 557–76.

Aronoff, Kate. "Could a Marshall Plan for the Planet Tackle the Climate Crisis?" *The Nation*, November 16, 2017. www.thenation.com. Accessed October 19, 2019.

Aronoff, Kate, Alyssa Battistoni, Daniel Aldana Cohen, and Thea Riofrancos. *A Planet to Win: Why We Need a Green New Deal*. New York: Verso, 2019.

Arseneault, C., and B. Pierson, eds. *Wings of Renewal: A Solarpunk Dragon Anthology*. Florida: Incandescent Phoenix Books, 2014.

Avila-Calero, Sofia. "Contesting Energy Transitions: Wind Power and Conflicts in the Isthmus of Tehuantepec." *Journal of Political Ecology* 24, no.1 (2017): 992–1012.

Bacevich, Andrew. *America's War for the Greater Middle East: A Military History*. New York: Random House, 2016.

Baiocchi, Gianpaolo. *We, the Sovereign*. Medford: Polity Press, 2018.

Bakke, Gretchen. *The Grid: The Fraying Wires Between Americans and Our Energy Future*. New York: Bloomsbury, 2016.

Barrett, Ross, and Daniel Worden, eds. *Oil Culture*. Minneapolis: University of Minnesota Press, 2014.

Bauman, Zygmunt. *Globalization: The Human Consequences*. New York: Columbia University Press, 1998.

Berners-Lee, Mike, and Duncan Clark. "What's the Carbon Footprint of . . . a New Car." *The Guardian*, September 23, 2010.

Blanchet, Cecile. "Is Renewable Energy a Commons?" *Resilience*, May 24, 2017.

Bond, Patrick. *Politics of Climate Justice: Paralysis Above, Movement Below.* Scottsville, South Africa: University of Kwazulu-Natal Press, 2012.

Bozuwa, Johanna, and Thomas M. Hanna. "Democratize Finance, Euthanize the Fossil Fuel Industry." *Jacobin*, March 7, 2019.

BP. "Two Steps Forward, One Step Back." June 13, 2018. Accessed October 19, 2019. www.bp.com.

Brennan, Shane. "Visionary Infrastructure: Community Solar Streetlights in Highland Park." *Journal of Visual Studies* 16, no. 2 (2017): 167–89.

Brown, D. Clayton, *Electricity for Rural America: The Fight for the REA.* Westport, CT: Greenwood Press, 1980.

Burnett, Victoria. "La Ventosa Journal: Mexico's Wind Farms Brought Prosperity, But Not for Everyone." *New York Times*, July 26, 2016.

Busch, Henner. Interview by author, February 13, 2017.

Caffentzis, George. "A Discourse on Prophetic Method: Oil Crises and Political Economy, Past and Future." in Kolya Abramsky, ed., *Sparking a Worldwide Energy Revolution: Social Struggles in the Transition to a Post-Petrol World.* Oakland, CA: AK Press, 2010.

Cagle, Susie. "Bees, Not Refugees: The Environmentalist Roots of Anti-Immigrant Bigotry." *Guardian*, August 16, 2019. www.theguardian.com. Accessed October 23, 2019.

Caiazzo, Fabio, et al. "Air Pollution and Early Deaths in the United States." *Atmospheric Environment* 79 (November 2013): 198–208.

Call, Henry L. *The Coming Revolution.* Boston: Arena Publishing, 1895.

Cardwell, Dianne. "Solar Experiment Lets Neighbors Trade Energy Among Themselves." *New York Times*, March 13, 2017.

Chen, Chung-Hwan. "Different Meanings of the Term Energeia in the Philosophy of Aristotle." *Philosophy and Phenomenological Research* 17, no. 1 (1956): 56-65.

Chen, Michelle. "Where Have All the Green Jobs Gone?" *The Nation*, April 22, 2014.

Chen, Stefanos. "At $238 Million, It's the Highest-Price Home in the Country." *New York Times,* January 23, 2019.

Christie, Jean. "Giant Power: A Progressive Proposal of the Nineteen Twenties." *Pennsylvania Magazine of History and Biography* 96, no. 1 (1972): 480–507.

Christie, Jean. *Morris Llewellyn Cooke, Progressive Engineer.* New York: Garland Publishing, 1983.

Cohen, Lizabeth. *Consumers' Republic: The Politics of Mass Consumption in Postwar America.* New York: Vintage, 2003.

Cooke to Pinchot, March 10, 1924. *Cooke Papers.* FDR Presidential Library and Museum, Box 35, File 391.

COP21: Shows the End of Fossil Fuels Is near, We Must Speed Its Coming. Greenpeace Philippines. https://www.greenpeace.org/archive-seasia/ph/News/greenpeace-philippine-blog/cop21-shows-the-end-of-fossil-fuels-is-near-w/blog/55098/. Accessed October 12, 2019.

Cox, Stan. *Losing Our Cool: Uncomfortable Truths about Our Air-Conditioned World.* New York: New Press, 2010.

Cox, Stan, and Paul Cox. "100% Wishful Thinking: The Green Energy Cornucopia." *Counterpunch,* September 14, 2017.

Counter Space: Design and the Modern Kitchen. September 15, 2010–May 2, 2011, the Museum of Modern Art, New York.

Cozzie, Laura, Olivia Chen, Aaron Koh, and Hannah Daly. "Commentary: Population Without Access to Electricity Falls Below 1 Billion." *International Energy Agency,* October 30, 2018, https://www.iea.org/newsroom/news/2018/october/population-without-access-to-electricity-falls-below-1-billion.html. Accessed October 23, 2019.

Crooks, Ed. "The US Shale Revolution: How It Changed the World (And Why Nothing Will Ever Be the Same Again)." *Financial Times,* April 24, 2015.

Cumbers, Andrew. "Public Ownership as Economic Democracy." in Cumbers, Andrew, et al., "Public Ownership and Alternative Political Horizons," *Soundings* 64 (Winter 2017): 83–104.

Cumbers, Andrew. *Reclaiming Public Ownership: Making Space for Economic Democracy.* New York: Zed Books, 2012.

Cumbers, Andrew. "Rethinking Public Ownership as Economic Democracy." *Alternatives to Neoliberalism: Towards Equality and Democracy*, 2017, pp. 209–26.

Daintith, Terence. "The Rule of Capture: The Least Worst Property Rule for Oil and Gas." In *Property and the Law in Energy and Natural Resources*, edited by Aileen McHarg, Barry Barton, Adrian Bradbrook, and Lee Godden. New York: Oxford University Press, 2010.

Dardot, Pierre, and Christian Laval. *Common: On Revolution in the 21st Century.* New York: Bloomsbury, 2019.

Dawson, Ashley. "Introduction: New Enclosures". *New formations* 69, no.1 (2009): 8–22.

Dawson, Ashley. *Extinction: A Radical History.* New York: O/R Books, 2016.

Democratic Socialists of America. "Con Ed and National Grid Talking Points and Research." Ecosocialist Working Group, July 22, 2019.

Dow, Alexander. "Exploding Some Myths on Superpower and "Giant Power." *NELA Bulletin XI*, 1925.

Dunlap, Alexander. "Wind Energy: Toward a 'Sustainable Violence" in Oaxaca." *NACAL Report on the Americas* 49, no. 4 (2017): 483–88.

Ediger, Volkan, and John Bowlus, "A Farewell to King Coal: Geopolitics, Energy Security, and the Transition to Oil, 1898-1917." *The Historical Journal* 62, no. 2 (2018): 1–23.

Ely, Richard. *Problems of To-Day: A Discussion of Protective Tariffs, Taxation, and Monopolies.* New York: Thomas Y. Crowell and Co., 1888.

Efler, Michael. "Personal Interview." February 3, 2017.

Ely, Richard. *An Introduction to Political Economy.* New York: Chattaqua Press, 1889.

Ely, Richard. *Outlines of Economics.* London: Macmillan, 1893.

Energy and Policy Institute. "Edison Electric Institute's Anti-Solar, PR Spending Revealed." January 21, 2015. https://www.energyandpolicy.org/edison-electric-institute-anti-solar-pr-spending-revealed/. Accessed October 18, 2019.

Energy Information Administration. *Short-Term Energy Outlook.* November 13, 2019.

Environment New York Research and Policy Center. "America's Dirtiest Power Plants." September 2014. www.environmentnewyorkcenter.org.

Environmental Justice Atlas. "Corporate Wind Farms in Ixtepec vs Community's Initiative, Oaxaca, Mexico." March 29, 2017. www.ejatlas.org. Accessed October 24, 2019.

"'Extraordinarily Hot' Arctic Temperatures Alarm Scientists." *Guardian*, November 22, 2016. https://www.theguardian.com/environment/2016/nov/22/extraordinarily-hot-arctic-temperatures-alarm-scientists. Accessed October 12, 2019.

Fairchild, Denise, and Al Weinrub. Eds. *Energy Democracy: Advancing Equity in Clean Energy Solutions.* Washington, DC: Island Press, 2017.

Federal Theatre Project. "Techniques Available to the Living Newspaper Dramatist." *Federal Theater Project booklet*, 1938.

Federal Theatre Project. *Federal Theatre Plays.* Edited by Pierre De Rohan. New York: Da Capo, 1973.

Federici, Silvia. *Re-enchanting the World: Feminism and the Politics of the Commons.* Oakland, CA: PM Press, 2019.

Fialka, John. "As Hawaii Aims for 100% Renewable Energy, Other States Watching Closely." *Scientific American*, April 27, 2018.

Figueres, Christiana, et al. "Three Years to Safeguard Our Climate." *Nature News* 546, no. 7660 (June 2017): 593. *www.nature.com*, doi:10.1038/546593a.

Finnegan, William. "Leasing the Rain," *New Yorker*, March 21, 2002.

Gallucci, Maria. "Energy Equity: Bringing Solar Power to Low-Income Communities." *Yale Environment 360*, April 4, 2019.

Gardiner, Beth. "For Europe's Far-Right Parties, Climate is the New Battleground." *Yale Environment 360*, October 29, 2019.

Goldberg, Noah. "Brooklynites Who Lost Power Ask Con Edison: "Why Us?" *Brooklyn Eagle*, July 22, 2019.

Gómez-Barris, Macarena .*The Extractive Zone: Social Ecologies and Decolonial Perspectives*. Durham, NC: Duke University Press, 2017.

Graeber, David. *Bullshit Jobs: A Theory*. New York: Simon & Schuster, 2018.

Greider, William. "Why the Federal Reserve Needs an Overhaul." *The Nation*, February 12, 2014.

Grunwald, Michael. *The New New Deal*. New York: Simon & Schuster, 2012.

Guse, Clayton. "Giant Electric Battery Set Will Help Curb Ravenswood Plant Pollution in Queens, State Says." *Daily News*, October 17, 2019.

Haigh, Jennifer. *Heat and Light*. New York: HarperCollins, 2016.

Hardin, Garrett. "The Tragedy of the Commons." *Science* 162, no. 3859 (1968): 1243–48.

Hardin, Garrett. "The Tragedy of the *Unmanaged* Commons: Population and the Disguises of Providence." In Robert V. Andelson, ed., *Commons Without Tragedy: Protecting the Environment from Overpopulation – A New Approach*. London: Shepheard-Walwyn, 1991.

Hardt, Michael, and Antonio Negri. *Commonwealth*. New York: Harvard University Press, 2009.

Harvey, David. *The New Imperialism*. New York: Oxford University Press, 2003.

Harvey, David. *Rebel Cities: From the Right to the City to the Urban Revolution*. New York: Verso, 2012.

Harvey, Fiona. "'Carbon Bubble' Could Spark Global Financial Crisis, Study Warns." *Guardian*, June 4, 2018.

Hayden, Dolores. *Building Suburbia: Green Fields and Urban Growth, 1820-2000*. New York: Vintage, 2004.

Hayek, Friedrich A. *The Road to Serfdom*, 2nd ed. Chicago: University of Chicago Press, 2007.

Hickel, Jason. "The Limits of Clean Energy." *Foreign Policy*, September 6, 2019.

Hiltner, Ken. *Forward to Nature: Writing a New Environmental Era*. New York: Routledge, 2019.

Hilton, Isabel. "How China's Belt and Road Initiative Threatens Global Climate Progress." *YaleE360*, January 3, 2019.

Holloway, John. *Change the World Without Taking Power*. New York, Pluto Press, 2002.

Honty, Gerardo. "Energy and Climate Change: No Progress in 20 Years." *Climate and Capitalism*, June 26, 2018. www.climateandcapitalism. com. Accessed October 19, 2019.

Howard Kunstler, James. *The Long Emergency: Surviving the Converging Crises of the Twenty-First Century*. New York: Grove/Atlantic, 2005.

Howe, Cymene. "Aeolian Extractivism and Community Wind in Southern Mexico." *Public Culture* 28, no. 2 (2016): 215–35.

Huber, Matthew. "Five Principles of a Socialist Climate Politics." *The Trouble*, August 16, 2018.

Huber, Matthew. *Lifeblood: Energy, Freedom, and the Forces of Capital*. Minnesota: University of Minnesota Press, 2013.

Huber, Matthew, Keith Brower Brown, Jeremy Gong, and Jamie Munro. "A Real Green New Deal Means Class Struggle." *Jacobin*, March 21, 2019. https://jacobinmag.com/2019/03/green-new-deal-class-struggle-organizing. Accessed October 28, 2019.

Hughes, Thomas P. *Networks of Power: Electrification in Western Society, 1880-1930*. Baltimore: Johns Hopkins University Press, 1983.

Hunter, Rob. "Waiting for SCOTUS." *Jacobin*, June 1, 2014. https://jacobinmag.com/2014/06/waiting-for-scotus. Accessed October 28, 2019.

Insull, Samuel. "President's Address." *National Electric Light Association, Proceedings*, 1898.

International Energy Agency. *World Energy Outlook 2017*, November 14, 2017. www.iea.org. Accessed October 19, 2019.

International Energy Agency. "Energy Access Outlook 2017." www.iea.org. Accessed October 19, 2019.

International Energy Agency. *Population Without Access to Electricity Falls Below 1 Billion*. October 30, 2018.

International Energy Agency. "Sustainable Development Scenario: A Cleaner and More Inclusive Energy Future." www.iea.org. Accessed October 19, 2019.

IRENA. "Economy and Human Welfare to Grow Under IRENA's 2050 Energy Transformation Road Map." www.irena.org. Accessed October 19, 2019.

IRENA. *30 Years of Policies for Wind Energy: Lessons from Denmark*. January 2013.

Johnson, Bob. *Carbon Nation: Fossil Fuels and the Making of American Culture*. Lawrence: University Press of Kansas, 2014.

Kahn, Brian. "We Just Breached the 410 PPM Threshold for CO_2: Carbon Dioxide Has Not Reached This Height In Millions of Years." *Scientific American*, April 21, 2017.

Kaempffert, Waldemar. "Power for the Abundant Life." *New York Times*, 23 Aug 1936.

Katsiaficas, George. *The Subversion of Politics: European Autonomous Social Movements and the Decolonization of Everyday Life*. Oakland, CA: AK Press, 2006.

Kilgannon, Corey. "Manhole Fires and Burst Pipes: How Winter Wreaks Havoc on What's Beneath NYC." *New York Times*, February 21, 2019.

Klein, Naomi. *The Shock Doctrine*. New York: Picador, 2007.

Klein, Naomi. *This Changes Everything: Capitalism vs The Climate* New York: Simon & Schuster, 2014.

Koeppel, Jason, Johanna Bozuwa, and Liz Veazey. "The Green New Deal Must Put Utilities Under Public Control." *In These Times*, February 1, 2019.

Koeppel, Jackson. "Organizing for Energy Democracy in the Face of Austerity." *Transnational Institute*, April 14, 2019.

Kulish , Nicolas, and Mike McIntire. "The New Nativists: Why an Heiress Spent Her Fortune Trying To Keep Immigrants Out." *New York Times*, August 14, 2019.

Leiden University. "Renewable Energy Sources Can Take Up to 1000 Times More Space Than Fossil Fuels." August 28, 2018. https://phys.org/news/2018-08-renewable-energy-sources-space-fossil.html. Accessed October 24, 2019.

Lennon, Myles. "Decolonizing Energy: Black Lives Matter and Technoscientific Expertise amid Solar Transitions." *Energy Research & Social Science* 30 (2017): 18–27. *onesearch.cuny.edu*, doi:10.1016/j.erss.2017.06.002.

Liebreich, Michael. "Trump's Influence on the Future of Clean Energy is Less Clear Than You Think." *The Guardian*, November 12, 2016.

Linebaugh, Peter. *The Magna Carta Manifesto: Liberties and Commons for All.* Berkeley: University of California Press, 2009.

London, Jennifer. "System Overload Slows Hawaii's Solar Energy Boom." *Al Jazeera*, January 10, 2014.

Lovins, Amory. "Resilience in Energy Strategy." *New York Times*, July 24, 1977.

Lovins, Amory. *Soft Energy Paths: Toward a More Durable Peace.* New York: HarperCollins, 1979.

Lloyd, H. D. "The Story of a Great Monopoly." *The Atlantic*, March 1881. https://www.theatlantic.com/magazine/archive/1881/03/the-story-of-a-great-monopoly/306019/. Accessed October 28, 2019.

Malm, Andrea. *The Progress of this Storm: Nature and Society in a Warming World.* New York: Verso, 2018.

Manly, Basil. "Editorial." *Pennsylvania Grange News*, August 1923.

Marchese, Anthony J., and Dan Zimmerle. "The US Natural Gas Industry is Leaking Way More Methane Than Previously Thought." *The Conversation*, July 2, 2018. www.theconversation.com. Accessed October 19, 2019.

Marois, Thomas. "How Public Banks Can Help Finance A Green and Just Energy Transition." *Public Alternatives Issue Brief.* Transnational Institute. November 2017.

Martinez, Cecilia. "From Commodification to the Commons: Charting the Pathway for Energy Democracy." In Fairchild, Denise, and Al Weinrub, Editors, *Energy Democracy: Advancing Equity in Clean Energy Solutions.* Washington, DC: Island Press, 2017.

Mazzucato, Mariana. *The Entrepreneurial State: Debunking Public vs. Private Sector Myths.* New York: PublicAffairs, 2015.

McGrath, Matt. "China Coal Power Building Boom Sparks Climate Warning." *BBC News*, September 26, 2018.

McKibben, Bill. "How Climate Activists Failed to Make Clear the Problem with Natural Gas." *Yale Environment 360*, March 13, 2018. www.e360. yale.edu. Accessed Ocotber 19, 2019.

McKibben, Bill. "Recalculating the Climate Math." *The New Republic*, September 22, 2016.

McKibben, Bill. "Power to the People: Why the Rise of Green Energy Makes Utility Companies Nervous." *The New Yorker*, June 22, 2015.

McLean, Bethany. "The Next Financial Crisis Lurks Underground: Fueled by Debt and Years of Easy Credit, America's Energy Boom is on Shaky Footing." *New York Times*, September 1, 2018.

McRae, Louise, et al. *Living Planet Report 2016: Risk and Resilience in a New Era.* 2016. http://awsassets.panda.org/downloads/lpr_2016_full_report_low_res.pdf. Accessed 12 Oct. 2019.

Mercure, J. F., et al. "Macroeconomic Impact of Stranded Fossil Fuel Assets." *Nature Climate Change*, June 4, 2018.

Mikulka, Justin. "GOP Tax Law Bails Out Fracking Companies Buried in Debt." *DeSmog Blog*, April 26, 2018.

Mikulka, Justin. "Peak Shale: Is the US Fracking Industry Already in Decline?" *DeSmog Blog*, October 30, 2018.

Mikulka, Justin. "Fracking in 2018: Another Year of Pretending to Make Money." *DeSmog Blog*, January 17, 2019.

Mill, John Stuart. *Principles of Political Economy, with Some of Their Applications to Social Philosophy.* London: Little, Brown, 1848.

Milun, Kathryn. *The Political Uncommons: The Cross-Cultural Logic of the Global Commons.* Burlington: Ashgate, 2011.

Milun, Kathryn, and Matthew Grimley. *Solar Commons: Designing Community Trust Solar Ownership for Social Equity,* 2017. www.solar-commons.org. Accessed October 23, 2019.

Mitchell, Timothy. *Carbon Democracy: Political Power in the Age of Oil.* New York: Verso, 2011.

Moore, Jason W. *Capitalism in the Web of Life: Ecology and the Accumulation of Capital.* New York: Verso, 2015.

More, Thomas. *Utopia.* New York: Adamant Media, 2005.

Morehouse, Catherine. "Chicago Considers Municipalizing ComEd." *Utilitydive,* July 25 2019.

Morris, Craig and Anre Jungjohann. *Energy Democracy: Germany's Energiewende to Renewables.* New York: Palgrave Macmillan, 2016.

Muehlebach, Andrea. "Towards a Social Infrastructure." *e-flux,* n.d.

Muttitt, Greg. "OFF TRACK: the IEA and Climate Change." *Oil Change International,* April 4, 2018.

National Rural Electric Co-operative Association. "2017 Fact Sheet." www.electric.co-op. Accessed October 10, 2019.

Neumann-Cosel, Luise. Interview by author, February 2, 2017.

New York City Department of Health. "Community Health Profiles 2015: Queens Community District 1." www1.nyc.gov.

ASHLEY DAWSON   227

New York Lawyers for the Public Interest. "We Must Stop New York's 'Peaker Plants' Choking Marginalized Communities." February 11, 2019. www.nylpi.org.

New York's Utility Project. "Borrowing at High Interest to Pay Unaffordable Utility Bills." October 10, 2012. www.utilityproject.org.

Nikiforuk, Andrew. *The Energy of Slaves: Oil and the New Servitude.* Vancouver: Greystone Books, 2012.

Nixon, Rob. *Slow Violence and the Environmentalism of the Poor.* New York: Harvard University Press, 2011.

Nixon, Rob. "Neoliberalism, Genre, and the 'Tragedy of the Commons." *PMLA* 127, no. 3 (2012): 593–99.

Nye, David E. *Consuming Power: A Social History of American Energies.* Cambridge, MA: MIT Press, 1999.

—. *Electrifying America: Social Meanings of a New Technology, 1880 – 1940.* Cambridge, MA: MIT Press, 1992.

Obama, Barack. "Remarks of Senator Barack Obama to the Detroit Economic Club." May 7, 2007.

Obama, Barack. "The Irreversible Momentum of Clean Energy." *Science* 355, no. 6321 (2017): 126–29.

O'Byrne, Ellie. "An Exploration of Oil as the Devil's Excrement." *Irish Times,* April 4, 2017.

Oceransky, Sergio. "Fighting the Enclosure of Wind: Indigenous Resistance to the Privitization of Wind Resources in Southern Mexico." In Kolya Abramsky, ed. *Sparking A Worldwide Energy Revolution Social Struggles In The Transition To A Post-Petrol World.* Oakland, CA: AK Press, 2010.

Office of Electricity. "Electricity 101." www.energy.gov. Accessed October 19, 2019.

Oil Change International. "The Sky's the Limit: Why the Paris Climate Goals Require a Managed Decline of Fossil Fuel Production." September 2016.

Oil Change International. "Fossil Fuel Subsidies Overview." www.pri-ceofoil.org. Accessed October 19, 2019.

Oil Change International. "G20 Energy Ministers Reaffirm Commitments to Fossil Gas, Compromising Paris Climate Commitment." June 18, 2018. www.priceofoil.org. Accessed October 19, 2019.

Olivera, Oscar. *!Cochabamba!: Water War in Bolivia*. New York: South End Press, 2004.

Olson, Bradley, Rebecca Elliot, and Christopher M. Matthews. "Fracking's Secret Problem – Oil Wells Aren't Producing as Much as Forecast." *Wall Street Journal*, January 2, 2019.

Osborne, James. "How Long Can the Fracking Spending Spree Last?" *Houston Chronicle*, September 14, 2018.

Ostrom, Elinor. *Governing the Commons: The Evolution of Institutions for Collective Action*. New York: Cambridge University Press, 1990.

Parr, Adrian. *The Wrath of Capital: Neoliberalism and Climate Change Politics.* New York: Columbia University Press, 2013.

Patel, Kiran Klaus. *The New Deal: A Global History.* Princeton, NJ: Princeton University Press, 2017.

Patterson, J., Wasserman, K., Starbuck, A., Sartor, A., Hatcher, J., Fleming, J., & Fink, K. *Coal Blooded: Putting Profits before People* 2011, https://www.naacp.org/climate-justice-resources/coal-blooded/. Accessed 19th Oct. 2019.

Pérez-Medina, Lourdes, and Elizabthe Yeampierre. "The People's Power." *Urban Omnibus*, April 10, 2019.

Peters, Glen. "Oil and Gas in a Low Carbon World." *Klima*, August 12, 2017. www.cicero.oslo.no. Accessed October 19, 2019.

Pinchot, Gifford. "Introduction." *The Annals of the American Academy of Political and Social Science* 118, no. 1 (1925): vii–xii.

Plaiss, Adam. "From Natural Monopoly to Public Utility: Technological Determinism and the Political Economy of Infrastructure in Progressive-Era America." *Technology and Culture* 57, no. 4 (2016): 806–30.

Plumer, Brad. "Here's Why 1.2 Billion People Still Don't Have Access to Electricity." *Washington Post*, May 29, 2013. https://www.washingtonpost.com/news/wonk/wp/2013/05/29/heres-why-1-2-billion-people-still-dont-have-access-to-electricity/. Accessed October 12, 2019.

Podobnik, Bruce. "Building the Clean Energy Movement: Future Possibilities in Historical Perspective." In Kolya Abramsky, ed. *Sparking a Worldwide Energy Revolution: Social Struggles in the Transition to a Post-Petrol World*. Oakland, CA: AK Press, 2010.

Polanyi, Karl. *The Great Transformation: The Political and Economic Origins of Our Time*. Boston, MA: Beacon Press, 1944.

Poulantzas, Nikos. *State, Power, Socialism*. New York: Verso, 2014.

Powalla, Oliver. "Personal Interview." February 23, 2018.

Prashad, Vijay. *The Darker Nations: A People's History of the Third World*. New York: New Press, 2008.

Public Power NYC. "Public Utilities Under Public Control." www.publicpower.nyc.

Rasch, Christopher. "Energiewende Retten." *Greenpeace Energy*, November 29, 2019.

Rayasam, Renuka. "A Power Grid of Their Own: German Village Becomes Model for Renewable Energy." *Der Spiegel*, March 9, 2012.

Renewable Energy Focus. "Onshore Wind Power Now as Affordable as Any Other Source, Solar to Halve by 2020." www.renewableenergyfocus.com. Accessed October 19, 2019.

REN21. "Renewable Energy Policy Network Renewables Global Status Report."

www.ren21.net. Accessed October 12, 2019.

REN21. "The State of the Global Renewable Energy Transition 2018." www.ren21.net. Accessed October 19, 2019.

Rewald, Rebecca. "Does Providing Energy Access Improve the Lives of Women and Girls? Sort of." *Oxfam*, June 6, 2017.

Roberts, David. "The International Energy Agency Consistently Underestimates Wind and Solar Power. Why?" *Vox*, October 12, 2015.

Rivoli, Dan. "Con Ed asks New Yorkers to Cough Up $695M in Rate Hikes." *Daily News*, January 31, 2019.

Roberts, David. "Shell's Vision of a Zero Carbon World by 2070, Explained." *Vox*, March 30, 2018.

Roberts, David. "The World's Bleak Climate Situation, in 3 Charts." *Vox*, May 1, 2018.

Roberts, David. "Clean Energy Technologies Threaten to Overwhelm the Grid. Here's How It Can Adapt." *Vox*, December 27, 2018.

Roberts, David. "The Story of Coal in the 21st Century, in One Amazing Map." *Vox*, June 7, 2018.

Roberts, David. "The Most Depressing Energy Chart of the Year." *Vox*, June 16, 2018. www.vox.com. Accessed October 19, 2019.

Roberts, David. "Sucking Carbon Out of the Air Won't Solve Climate Change." *Vox*, July 16, 2018.

Roberts, David. "Energy Lobbyists Have a New PAC to Push for A Carbon Tax." *Vox*, June 23, 2018.

Rocholl, Nora, and Ronan Bolton. "Berlin's Electricity Distribution Grid: An Urban Energy Transition in a National Regulatory Context." *Technology Analysis and Strategic Management* 28, no. 10 (2016): 1188.

Royal Dutch Shell. *The Sky Scenario*. www.shell.com. Accessed October 19, 2019.

Rudolph, Richard, and Scott Ridley. *Power Struggle: The Hundred-Year War over Electricity*. New York: Harper & Row, 1986.

Ruivenkamp, Guido, and Andy Hilton. "Introduction." In *Perspectives on Commoning: Autonomist Principles and Practices*, edited by Guido Ruivenkamp and Andy Hilton. London: Zed Books, 2017.

Ruther, Amber, and Steve Fox. "Personal Interview." August 15, 2019.

Salminen, Antti, and Tere Vadén. *Energy and Experience: An Essay in Nafthology.* MCM, 2015.

Sammon, Alexander. "How the Bank Bailout Hobbled the Climate Fight." *New Republic*, October 22, 2018.

Sassen, Saskia. *Expulsions: Brutality and Complexity in the Global Economy.* Cambridge, MA: Belknap Press, 2014.

Sayer, Andrew. *Why We Can't Afford the Rich.* Chicago: University of Chicago Press, 2014.

"Scientists Warn World Will Miss Key Climate Target." *Guardian*, August 6, 2016. https://www.theguardian.com/science/2016/aug/06/global-warming-target-miss-scientists-warn. Accessed October 12, 2019.

Schneider, Nathan. "Energy Democracy and the Billion-Dollar Co-Op." *The Nation*, May 8, 2017.

Schultz, Jim. "The Politics of Water in Bolivia." *The Nation*, January 28, 2005.

Schuur, E. A. G., et al. "Climate Change and the Permafrost Carbon Feedback." *Nature* 520, no. 1 (2015): 171–79.

Sengupta, Somini. "Why Build Kenya's First Coal Plant? Hint: Think China." *New York Times*, February 27, 2018.

Sheats, Nicky. "Achieving Emissions Reductions for Environmental Justice Communities Through Climate Change Mitigation Policy." *William and Mary Environmental Law and Policy Review* 377, no. 1 (2017): 376–402.

Sitrin, Marina, ed. *Horizontalism: Voices of Popular Power in Argentina.* Oakland, CA: AK Press, 2006.

Slezak, Michael. "Asia's Coal-Fired Power Boom Bankrolled by Foreign Governments and Banks." *Guardian*, July 20, 2017.

Smil, Vaclav. *Power Density: A Key to Understanding Energy Sources and Uses.* Cambridge, MA: MIT Press, 2015.

Smith, Matthew Nitch. "The Number of Cars Will Double Worldwide by 2040." *Business Insider,* April 20, 2016.

Smith, Richard. "Capitalism and the Destruction of Life in Earth: Six Theses on Saving the Humans." *Real-World Economics Review,* 2013.

"Solar Commons: How Does It Work?" *The Solar Commons.* www.solar-commons.org. Accessed October 23, 2019.

Southern Poverty Law Center. "Garrett Hardin." www.splcenter.org. Accessed October 24, 2019.

Spiegel, Jan Ellen. "Another $1.2 billion substation? No thanks, says utility, we'll find a better way." *Inside Climate News,* April 4, 2016.

Stark, Kevin. "Power Switch: SF Builds Case for Pushing Out PG&E." *SFPublic Press,* June 18, 2019.

Stern, Nicholas. *The Economics of Climate Change: The Stern Review.* New York: Cambridge University Press, 2007.

Sweeney, Sean, and John Treat. "Are We Winning?" *Trade Unions for Energy Democracy.* Working Paper #9, January 2017.

Sweeney, Sean, and John Treat, "Preparing a Public Pathway: Confronting the Investment Crisis in Renewable Energy." Trade Unions for Energy Democracy Working Paper #10, November 2017.

Szeman, Imre. *After Oil.* Petrocultures Research Group, 2016.

Szeman, Imre, Jennifer Wenzel, and Patricia Yaeger, eds. *Fueling Culture: 101 Words for Energy and Environment.* New York: Fordham University Press, 2017.

"The '100% Renewable' Dream Will Require Producing Less Energy and More Energy Justice." *Green Social Thought.* http://greensocialthought.org/content/100-renewable-dream-will-require-producing-less-energy-and-more-energy-justice. Accessed October 12, 2019.

"The Global Climate in 2011–2015." *World Meteorological Organization,* November 8, 2016, https://public.wmo.int/en/resources/library/global-climate-2011%E2%80%932015. Accessed October 19, 2019.

The Breakthrough Institute. "Ecomodernist Manifesto." April 2015. www.ecomodernism.org. Accessed October 19, 2019.

Thompson, Carl D. *Confessions of The Power Trust: A Summary of the Testimony Given in the Hearings of the Federal Trade Commission on Utility Corporations Pursuant to Resolution No. 83 of the Unites States Senate.* New York: E. P. Dutton, 1932.

Thompson, Helen. *Oil and the Western Economic Crisis.* Cambridge: Springer, 2017.

Tollefson, Jeff. "Can the World Kick its Fossil Fuel Addiction Fast Enough?" *Nature,* April 25, 2018.

Torralba, Alanah, Tadzio Müller, and Elis Soldatelli. "Pumping the Brakes on E-cars: Unmasking the Fantasy of Green Capitalism." *Rosa Luxemburg Stiftung,* December 11, 2019.

Transnational Institute. *Reclaiming Public Services: How Cities and Citizens Are Turning Back Privatization.* June 23, 2017.

Tschinkel, Sheila. "The Federal Reserve Needs to Remain Independent on the Whims of Politicians." *The Conversation,* July 23, 2018.

"TUED Working Paper #9: Energy Transition: Are We 'Winning'?" *Trade Unions for Energy Democracy,* 2017. http://unionsforenergydemocracy.org/resources/tued-publications/tued-working-paper-9-energy-transition-are-we-winning/. Accessed October 12, 2019.

Union of Concerned Scientists. "Environmental Impacts of Natural Gas." June 19 2014.

United Nations General Assembly Document. *Declaration for the Establishment of a New International Economic Order,* General Assembly Resolution 3201 (S-VI) of 1 May 1974. Published by the United Nations Audiovisual Library of International Law.

United Nations. "Sustainable Development Knowledge Platform: Sustainable Development Goal 7 – Ensure Access to Affordable, Reliable, Sustainable and Modern Energy for All." https://sustainabledevelopment.un.org/sdg7. Accessed October 19, 2019.

US Energy Information Administration. "US Electric System is Made up of Interconnections and Balancing Authorities." July 20, 2016. https://www.eia.gov/todayinenergy/detail.php?id=27152. Accessed October 19, 2019.

Vaidyanathan, Gayathri. "How Bad of a Greenhouse Gas is Methane?" *Scientific American*, December 22, 2015.

Vamburkar, Meenal. "Actions of Indicted ex-energy CEO killed in car crash aren't uncommon across US shale patch." *Chicago Tribune*, March 4, 2016.

Van Zalk, John, and Paul Behrens. "The Spatial Extent of Renewable and Non-Renewable Power Generation." *Energy Policy* 123, no. 1 (2018): 83–91.

Veltmeyer, Henry, and J. Petras. *The New Extractivism: A Post-Neoliberal Development Model or Imperialism of the 21st Century?* London: Zed Books, 2014.

Wade, Will, Brian Eckhouse, and Henry Goldman. "NYC Mayor Suggests ConEd Takeover After Heat Forces Shutdown." *Bloomberg*, July 21, 2019.

Wald, Matthew L. "The Blackout that Exposed the Flaws in the Grid." *New York Times*, November 11, 2013.

Watson, T. X. "The Boston Hearth Project." In *Sunvault: Stories of Solarpunk and Eco-Speculation*, edited by Phoebe Wagner and Brontë Christopher Wieland. Nashville, NT: Upper Rubber Boot Books, 2017.

What Is US Electricity Generation by Energy Source? - FAQ - US Energy Information Administration (EIA). https://www.eia.gov/tools/faqs/faq.php?id=427&t=3. Accessed October 12, 2019.

White, Richard. "For Tech Giants, a Cautionary Tale from 19th Century Railroads on the Limits of Competition." *The Conversation*, March 6, 2018.

"Who We Are." *The Solar Commons.* www.solarcommons.org. Accessed October 28, 2019.

Williams, Raymond. *Marxism and Culture.* New York: Oxford University Press, 1977.

Williams, Rhys. "This Shining Confluence of Magic and Technology: Solarpunk, Energy Imaginaries, and the Infrastructures of Solarity." *Open Library of Humanities* 5, no. 1 (2019): 2–35.

Wilson, Sheena, Adam Carlson, and Imre Szeman, eds. *Petrocultures: Oil, Politics, Culture.* Montreal: McGill-Queen's University Press, 2017.

World Bank. *The Growing Role of Minerals and Metals for a Low Carbon Future.* June 2017.

World Energy Outlook 2016. https://www.iea.org/newsroom/news/2016/november/world-energy-outlook-2016. Accessed October 12, 2019.

Zárate-Toledo, Ezequiel, Rodrigo Patiño, and Julia Fraga. "Justice, Social Exclusion and Indigenous Opposition: A Case Study of Wind Energy Development on the Isthmus of Tehuantepec, Mexico." *Energy Research & Social Science* 54, no. 1 (2019): 1–11.

Zuboff, Shoshona. *The Age of Surveillance Capitalism: The Fight for a Human Future at the New Frontier of Power.* New York: PublicAffairs, 2019.

NOTES

1. On the Capitalocene, see Jason W. Moore, *Capitalism in the Web of Life* (New York: Verso), 169–92.

2. World Meteorological Association, *The Global Climate 2011-2015* (November 8, 2016)

3. Robin McKie, "Scientists Warn World Will Miss Key Climate Target," *Guardian*, August 6, 2016.

4. John Vidal, "'Extraordinarily Hot' Arctic Temperatures Alarm Scientists," *Guardian*, November 22, 2016.

5. WWF, *Living Planet Report 2016: Risk and Resilience in a New Era.*

6. International Energy Association (IEA), *World Energy Outlook 2016*, 1.

7. On extractivism, see Henry Veltmeyer and James Petras, *The New Extractivism: A Post-Neoliberal Development Model or Neoliberalism in the Twenty-First Century* (London: Zed Books, 2014); Saskia Sassen, *Expulsions: Brutality and Complexity in the Global Economy* (Cambridge, MA: Belknap Press, 2014); and Macarena Gómez-Barris, *The Extractive Zone: Social Ecologies and Decolonial Perspectives* (Durham, NC: Duke University Press, 2017).

8. Ibid.

9. Bill McKibben, "Recalculating the Climate Math: The Numbers on Global Warming Are Even Scarier than We Thought," *New Republic*, September 22, 2016.

10. McKibben, "Recalculating the Climate Math: The Numbers on Global Warming Are Even Scarier than We Thought."

11. Petrocultures Research Group, *After Oil* (Petrocultures Research Group, 2016), 10.

12. Petrocultures Research Group, *After Oil*, 9.

13. The concept of "structures of feeling" was developed by Raymond Williams in *Marxism and Culture* (Oxford University Press, 1977).

14. On globalization and space/time compression, see Zygmunt Bauman, *Globalization: The Human Consequences* (New York: Columbia University Press, 1998).

15. The developing field of energy humanities seeks to explore the cultural lineaments of carbon culture. For representative examples of this work, see Sheena Wilson, Adam Carlson, and Imre Szeman, eds., *Petrocultures: Oil, Politics, Culture* (Montreal: McGill-Queen's University Press, 2017); Imre Szeman, Jennifer Wenzel, and Patricia Yaeger, eds., *Fueling Culture: 101 Words for Energy and Environment* (New York: Fordham University Press, 2017); and Ross Barrett and Daniel Worden, eds., *Oil Culture* (Minneapolis: University of Minnesota Press, 2014).

16. Kumi Naidoo, "COP21: shows the end of fossil fuels is near, we must speed its coming," *Greenpeace*, December 12, 2015.

17. That figure excludes 7.5 percent traditional hydropower. See US Energy Information Administration, "What is US Energy Generation By Source?," https://www.eia.gov/tools/faqs/faq.php?id=427&t=3.

18. Christiana Figueres et al., "Three Years to Safeguard Our Planet," *Nature* 546, no. 7660 (June 28, 2017).

19. Ibid.

20. Sean Sweeney and John Treat, "Energy Transition: Are We Winning? (Working Paper #9)," *Trade Unions for Energy Democracy* (January 2017).

21. My arguments here are drawn from Sweeney and Treat's excellent work.

22. REN21, "Estimated Renewable Energy Share of Global Final Energy Consumption, 2014," *Renewables 2016 Global Status Report*, www.ren21.net.

23. REN21, "Estimated Renewable Energy Share."

24. On the mass extinction crisis, see my book *Extinction: A Radical History* (O/R Books, 2016).

25. Labor Network for Sustainability, "'Just Transition' –What Is it?," www.ecology.iww.org.

26. On energy as a vital part of the global commons, see Kolya Abramsky, "Energy, Work, and Social Reproduction in the World-Economy," in Kolya Abramsky, ed., *Sparking a Worldwide Energy Revolution: Social*

Struggles in the Transition to a Post-Petrol World (Oakland, CA: AK Press, 2010), 93–101.

27. Cecile Blanchet, "Is Renewable Energy a Commons?" *Resilience.org*, May 24, 2017.

28. "Energy," *Oxford English Dictionary.*

29. Chung-Hwan Chen, "Different Meanings of the Term *Energeia* in the Philosophy of Aristotle," *Philosophy and Phenomenological Research* 17, no. 1 (September 1956): 56–65.

30. On slavery and energy, see Andrew Nikiforuk, *The Energy of Slaves: Oil and the New Servitude* (Vancouver: Greystone Books, 2012); and Miles Lennon, "Decolonizing Energy: Black Lives Matter and Technoscientific Expertise amid Solar Transitions," *Energy Research and Social Science* 30 (2017): 18–27.

31. On Blockadia, see Naomi Klein, *This Changes Everything: Capitalism vs The Climate* (New York: Simon & Schuster, 2014), 295–365.

32. Stan Cox, "The '100% Renewable' Dream Will Require Producing Less Energy and More Energy Justice," *Green Social Thought*, November 13, 2017.

33. Abramsky, 11.

34. See, for example, Friedrich Hayek, *The Road to Serfdom.* For a very useful discussion of public ownership, see Andrew Cumbers, "Public Ownership as Economic Democracy."

35. Naomi Klein, *The Shock Doctrine.*

36. Andrew Cumbers, "Public Ownership as Economic Democracy," 84.

37. Fairchild, *Energy Democracy*, 12.

38. Brad Plumer, "Here's Why 1.2 Billion People Still Don't Have Access to Electricity," *The Washington Post*, April 13, 2018.

39. See Nikos Poulantzas, *State, Power, Socialism* (New York: Verso), 256.

40. Barack Obama, "Remarks of Senator Barack Obama to the Detroit Economic Club" (May 7, 2007).

41. Michelle Chen, "Where Have All the Green Jobs Gone?" *The Nation*, April 22, 2014.

42. Michael Grunwald, *The New New Deal* (New York: Simon & Schuster, 2012), 360.

43. Chen, "Where Have All the Green Jobs Gone?"

44. Chen.

45. Jeremy Brecher, quoted in Chen.

46. Chen.

47. Ben Adler, "How Are Environmentalists Reacting to Obama's Clean Power Plan?," *Grist*, August 3, 2015.

48. Nicky Sheats, "Achieving Emissions Reductions for Environmental Justice Communities Through Climate Change Mitigation Policy" *William and Mary Environmental Law and Policy Review* 377 (2017): 377–402.

49. Barack Obama, "The Irreversible Momentum of Clean Energy," *Science* 355, no. 6321 (January 13, 2017): 126–29.

50. Obama, "Irreversible Momentum."

51. Michael Liebreich, "Trump's Influence on the Future of Clean Energy is Less Clear Than You Think," *Guardian*, November 12, 2016.

52. Ibid.

53. Christiana Figueres, quoted in Fiona Harvey, "'Carbon Bubble' Could Spark Global Financial Crisis, Study Warns," *Guardian*, June 4, 2018.

54. Sean Sweeney and John Treat, *Working Paper #9: Are We Winning? Trade Unions for Energy Democracy* (January 2017), 2.

55. On the history of the consumer nation, see Lizabeth Cohen, *Consumers' Republic: The Politics of Mass Consumption in Postwar America* (New York: Vintage, 2003).

56. Nicholas Stern, *The Economics of Climate Change: The Stern Review* (New York: Cambridge University Press, 2007).

57. Karl Polanyi, *The Great Transformation: The Political and Economic Origins of Our Time* (Boston, MA: Beacon Press, 1944), 3.

58. Polanyi, 147.

59. Polanyi, 146.

60. Mariana Mazzucato, *The Entrepreneurial State: Debunking Public vs. Private Sector Myths* (PublicAffairs, 2015).

61. Oil Change International, "Fossil Fuel Subsidies Overview," www .priceofoil.org.

62. Ibid.

63. Jennifer London, "System Overload Slows Hawaii's Solar Energy Boom," *Al Jazeera* (January 10, 2014).

64. London, "System Overload."

65. John Fialka, "As Hawaii Aims for 100% Renewable Energy, Other States Watching Closely," *Scientific American*, April 27 2018.

66. Fialka, "As Hawaii Aims."

67. "Electricity 101," Office of Electricity, www.energy.gov.

68. America's Electric Co-ops, "2017 Fact Sheet," www.electric.co-op.

69. US Energy Information Administration, "Today in Energy" (July 20, 2016).

70. Matthew L. Wald, "The Blackout that Exposed the Flaws in the Grid," *New York Times*, November 11, 2013.

71. David Roberts, "Clean Energy Technologies Threaten to Overwhelm the Grid. Here's How It Can Adapt," *Vox*, December 27, 2018.

72. International Energy Agency, *Energy Access Outlook 2017*, www.iea.org.

73. Sabrina Naz et al., "Household Air Pollution from Use of Cooking Fuel and Under-Five Mortality," PLoS ONE 12, no. 3 (2017): e0173256.

74. On the fossil underpinnings of the American way of life, see Matt Huber, *Lifeblood: Energy, Freedom, and the Forces of Capital* (Minneapolis: University of Minnesota Press, 2013).

75. Stan Cox, *Losing Our Cool: Uncomfortable Truths about Our Air-Conditioned World* (New York: New Press, 2010).

76. Huber, xiv.

77. Amory Lovins, "Resilience in Energy Strategy," *New York Times* (July 24, 1977).

78. Amory Lovins, *Soft Energy Paths: Toward a More Durable Peace* (HarperCollins, 1979).

79. J. Patterson, K. Wasserman, A. Starbuck, A. Sartor, J. Hatcher, J. Fleming, and K. Fink. *Coal Blooded: Putting Profits before People* (2011), < https://www.naacp.org/climate-justice-resources/coal-blooded/ >.

80. Bill McKibben, "Power to the People: Why the Rise of Green Energy Makes Utility Companies Nervous," *New Yorker*, June 22, 2015.

81. Soren Amelang, "Battered Utilities Take on Start Ups in Innovation Race," *Clean Energy Wire*, May 16, 2017.

82. Amelang, "Battered Utilities."

83. Energy and Policy Institute, "Edison Electric Institute's Anti-Solar, PR Spending Revealed," January 21, 2015.

84. Quoted in Jason Koeppel, Johanna Bozuwa, and Liz Veazey, "The Green New Deal Must Put Utilities Under Public Control," *In These Times*, February 1, 2019.

85. International Energy Agency, *World Energy Outlook 2017*, November 14, 2017, www.iea.org.

86. Sean Sweeney and John Treat, *Energy Transition*, 8.

87. David Roberts, "The Story of Coal in the 21st Century, in One Amazing Map," *Vox*, June 7, 2018.

88. Michael Slezak, "Asia's Coal-Fired Power Boom Bankrolled by Foreign Governments and Banks," *Guardian*, July 20, 2017.

89. David Roberts, "The World's Bleak Climate Situation, in 3 Charts," *Vox*, May 1, 2018.

90. Greg Muttitt, "OFF TRACK: the IEA and Climate Change," *Oil Change International*, April 4, 2018.

91. REN21, *The State of the Global Renewable Energy Transition 2018*.

92. Jeff Tollefson, "Can the World Kick its Fossil Fuel Addiction Fast Enough?," *Nature* 556, no. 7702 (April 25, 2018): 422-425.

93. Oil Change International, *The Sky's the Limit: Why the Paris Climate Goals Require a Managed Decline of Fossil Fuel Production*, September 2016.

94. Glen Peters, "Oil and Gas in a Low Carbon World," *Klima*, August 12, 2017, www.cicero.oslo.no.

95. David Roberts, "Shell's Vision of a Zero Carbon World by 2070, Explained," *Vox*, March 30, 2018.

96. Royal Dutch Shell, *The Sky Scenario*, www.shell.com.

97. Roberts, "Shell's Vision."

98. Kate Aronoff, "Could a Marshall Plan for the Planet Tackle the Climate Crisis?," *The Nation*, November 16, 2017, www.thenation.com.

99. David Roberts, "Sucking Carbon Out of the Air Won't Solve Climate Change," *Vox*, July 16, 2018, www.vox.com.

100. Roberts, "Shell's Vision."

101. Ibid.

102. E. A. G. Schuur et al., "Climate Change and the Permafrost Carbon Feedback," *Nature* 520 (April 9, 2015): 171–79.

103. REN21, "Renewables 2018: Global Status Report," www.ren21.net.

104. Ibid.

105. "Onshore Wind Power Now as Affordable as Any Other Source, Solar to Halve by 2020," *Renewable Energy Focus*, www.renewableenergyfocus. com.

106. J.-F. Mercure et al., "Macroeconomic Impact of Stranded Fossil Fuel Assets," *Nature Climate Change* 8 (June 4, 2018): 588-593.

107. Quoted in Fiona Harvey, "'Carbon Bubble' Could Spark Global Financial Crisis, Study Warns," *The Guardian* (June 4, 2018).]

108. See, for instance, Sean Sweeney and John Treat, *Energy Transition: Are We Winning?* (Trade Unions for Energy Democracy Working Paper #9, January 2017) www.unionsforenergydemocracy.org.

109. BP, "Two Steps Forward, One Step Back," June 13, 2018, www.bp.com.

110. Quoted in Gerardo Honty, "Energy and Climate Change: No Progress in 20 Years," *Climate and Capitalism* (June 26, 2018), www .climateandcapitalism.com

111. Matt McGrath, "China Coal Power Building Boom Sparks Climate Warning," *BBC News*, September 26, 2018.

112. Somini Sengupta, "Why Build Kenya's First Coal Plant? Hint: Think China." *New York Times*, February 27, 2018.

113. David Roberts, "The Most Depressing Energy Chart of the Year," *Vox*, June 16, 2018, www.vox.com.

114. International Energy Association, "Sustainable Development Scenario: A Cleaner and More Inclusive Energy Future," www.iea.org.

115. David Roberts, "The International Energy Agency Consistently Underestimates Wind and Solar Power. Why?," *Vox*, October 12, 2015.

116. United Nations, *Sustainable Development Knowledge Platform: Sustainable Development Goal 7 – Ensure Access to Affordable, Reliable, Sustainable and Modern Energy for All*, https://sustainabledevelopment.un.org.

117. IRENA, "Economy and Human Welfare to Grow Under IRENA's 2050 Energy Transformation Road Map," www.irena.org.

118. REN21, *The State of the Global Renewable Energy Transition 2018*, www .ren21.net.

119. Ibid.

120. Ibid.

121. Barack Obama, "The Irreversible Momentum of Clean Energy," *Science* 355, no. 6321 (Janaury 13, 2017).

122. Nate Aden, "The Roads to Decoupling: 21 Countries Are Reducing Carbon Emissions While Growing GDP," World Resources Institute, April 5, 2016.

123. The Breakthrough Institute, *Ecomodernist Manifesto* (April 2015), www.ecomodernism.org.

124. Ibid.

125. Adrian Parr, *The Wrath of Capital: Neoliberalism and Climate Change Politics* (New York: Columbia University Press, 2013), 27.

126. Patrick Bond, *Politics of Climate Justice: Paralysis Above, Movement Below* (Scottsville, South Africa: University of Kwazulu-Natal Press, 2012), 32.

127. Bond, *Politics of Climate Justice*, 31.

128. David Roberts, "Energy Lobbyists Have a New PAC to Push for A Carbon Tax," *Vox*, June 23, 2018.

129. REN21, *The State of the Global Renewable Energy Transition 2018*.

130. BP, "Two Steps Forward, One Step Back."

131. Sean Sweeney and John Treat, *Energy Transition: Are We Winning—Trade Unions for Energy Democracy, Working Paper #9*, January 2017, www.unionsforenergydemocracy.org.

132. Brian Kahn, "We Just Breached the 410 PPM Threshold for CO2: Carbon Dioxide Has Not Reached This Height In Millions of Years," *Scientific American*, April 21, 2017, https://www.scientificamerican.com/article/we-just-breached-the-410-ppm-threshold-for-co2/?WT.mc_id=SA_TW_ENGYSUS_NEWS.

133. Ibid.

134. Oil Change International, *G20 Energy Ministers Reaffirm Commitments to Fossil Gas, Compromising Paris Climate Commitment* (June 18, 2018), www.priceofoil.org.

135. Bill McKibben, "How Climate Activists Failed to Make Clear the Problem with Natural Gas," *Yale Environment 360*, March 13, 2018, www.e360.yale.edu.

136. McKibben, "How Climate Activists Failed."

137. Gayathri Vaidyanathan, "How Bad of a Greenhouse Gas is Methane?," *Scientific American*, December 22, 2015, https://www.scientificamerican.com/article/how-bad-of-a-greenhouse-gas-is-methane/

138. Anthony J. Marchese and Dan Zimmerle, "The US Natural Gas Industry is Leaking Way More Methane Than Previously Thought," *The Conversation*, July 2, 2018, www.theconversation.com.

139. McKibben, How Climate Activists Failed."

140. Jennifer Haigh, *Heat and Light* (New York: HarperCollins, 2016), 27, 30.

141. Haigh, 29.

142. On these earlier novels and the oil encounter, see Rob Nixon, *Slow Violence and the Environmentalism of the Poor* (Cambridge, MA: Harvard University Press, 2011), 86–89.

143. Haigh, 401.

144. Haigh, 402.

145. Meenal Vamburkar, "Actions of Indicted ex-energy CEO killed in car crash aren't uncommon across US shale patch," *Chicago Tribune*, March 4, 2016.

146. Ed Crooks, "The US Shale Revolution: How It Changed the World (And Why Nothing Will Ever Be the Same Again)," *Financial Times*, April 24, 2015.

147. Crooks, "The US Shale Revolution."

148. James Osborne, "How Long Can the Fracking Spending Spree Last?," *Houston Chronicle*, September 14, 2018.

149. Alexander Sammon, "How the Bank Bailout Hobbled the Climate Fight," *New Republic*, October 22, 2018.

150. James Howard Kunstler, *The Long Emergency: Surviving the Converging Crises of the Twenty-First Century* (New York: Grove/Atlantic, 2005).

151. Sammon, "How The Bank Bailout Hobbled the Climate Fight."

152. Helen Thompson, *Oil and the Western Economic Crisis* (New York: Springer, 2017).

153. Sammon, "How The Bank Bailout Hobbled the Climate Fight."

154. Sammon, "How The Bank Bailout Hobbled the Climate Fight."

155. Bradley Olson, Rebecca Elliot, and Christopher M. Matthews, "Fracking's Secret Problem—Oil Wells Aren't Producing as Much as Forecast," *Wall Street Journal*, January 2, 2019.

156. Justin Mikulka, "Peak Shale: Is the US Fracking Industry Already in Decline?," *DeSmog Blog*, October 30, 2018.

157. Bethany McLean, "The Next Financial Crisis Lurks Underground: Fueled by Debt and Years of Easy Credit, America's Energy Boom is on Shaky Footing," *New York Times*, September 1, 2018.

158. Justin Mikulka, "Fracking in 2018: Another Year of Pretending to Make Money," *DeSmog Blog*, January 17, 2019.

159. Justin Mikulka, "GOP Tax Law Bails Out Fracking Companies Buried in Debt," *DeSmog Blog*, April 26, 2018.

160. Justin Mikulka, "GOP Tax Law Bails Out Fracking Companies Buried in Debt."

161. Johanna Bozuwa and Thomas M. Hanna, "Democratize Finance, Euthanize the Fossil Fuel Industry," *Jacobin*, March 7, 2019.

162. See, for example, Sheila Tschinkel, "The Federal Reserve Needs to Remain Independent on the Whims of Politicians," *The Conversation*, July 23, 2018.

163. William Greider, "Why the Federal Reserve Needs an Overhaul," *The Nation*, February 12, 2014.

164. Thomas Marois, "How Public Banks Can Help Finance A Green and Just Energy Transition," *Public Alternatives Issue Brief*, Transnational Institute, November 2017.

165. Marois, 3.

166. Bozuwa and Hanna, "Democratize Finance."

167. Ibid.

168. Gar Alperovitz and Johanna Bozuwa, "Electric Companies Won't Go Green Unless the Public Takes Control," *In These Times*, April 22, 2019.

169. Emily Holden, "How the Oil Industry Has Spent Billions to Control the Climate Change Conversation," *Guardian*, January 8, 2020.

170. Gar Alperovitz, Joe Guinan, and Thomas M. Hanna, "The Policy Weapon Climate Activists Need," *The Nation*, April 26, 2017.

171. Although I am highly indebted to Bob Johnson for his excavation of debates around Giant Power and electrification during the New Deal, my interpretations of these debates differ in significant ways. See Bob Johnson, *Carbon Nation: Fossil Fuels and the Making of American Culture* (Lawrence: University Press of Kansas, 2014).

172. On the history of the "Giant Power" plan, see Jean Christie, "Giant Power: A Progressive Proposal of the Nineteen Twenties," *Pennsylvania Magazine*

of History and Biography 96 (October 1972): 480–507; Jean Christie, *Morris Llewellyn Cooke, Progressive Engineer* (New York: Garland Publishing, 1983); and Thomas P. Hughes, *Networks of Power: Electrification in Western Society, 1880-1930* (Baltimore: Johns Hopkins University Press, 1983).

173. Gifford Pinchot, "Introduction," *The Annals of the American Academy of Political and Social Science*, vol. 118, (March 1925), vii–xii.

174. Pinchot, vii.

175. Pinchot, vii.

176. Pinchot, viii.

177. Kolya Abramsky, "Introduction: Racing to 'Save' the Economy and the Planet: Capitalist or Post-Capitalist Transition to a Post-Petrol World?" in Kolya Abramsky, ed., *Sparking a Worldwide Energy Revolution: Social Struggles in the Transition to a Post-Petrol World* (Oakland, CA: AK Press, 2010), 8.

178. Timothy Mitchell, *Carbon Democracy: Political Power in the Age of Oil* (New York: Verso, 2011), 15.

179. Bruce Podobnik, "Building the Clean Energy Movement: Future Possibilities in Historical Perspective" in Kolya Abramsky, ed., *Sparking a Worldwide Energy Revolution: Social Struggles in the Transition to a Post-Petrol World* (Oakland, CA: AK Press, 2010), 72–90.

180. Podobnik, 74.

181. Volkan Ediger and John Bowlus, "A Farewell to King Coal: Geopolitics, Energy Security, and the Transition to Oil, 1898-1917," *The Historical Journal* 62, no. 2 (2018): 1–23.

182. Mitchell, 12–28.

183. On the role of petroleum's specific material characteristics in defeating militant labor, see Mitchell, 36–38.

184. Podobnik, 75.

185. On the Tricontinental Conference, see Vijay Prashad, *The Darker Nations: A People's History of the Third World* (New York: New Press, 2008), 105–18.

186. *Declaration for the Establishment of a New International Economic Order*, United Nations General Assembly Document A/RES/S-6/3201 of 1 May 1974.

187. George Caffentzis, "A Discourse on Prophetic Method: Oil Crises and Political Economy, Past and Future" in Kolya Abramsky, ed., *Sparking a*

Worldwide Energy Revolution: Social Struggles in the Transition to a Post-Petrol World (Oakland, CA: AK Press, 2010), 61–71. On the energy crisis, Big Oil, and petroleum price increases, see also Mitchell, *Carbon Democracy*, 178–92.

188. Caffentzis, 65.

189. Andrew Bacevich, *America's War for the Greater Middle East: A Military History* (New York: Random House, 2016).

190. Pinchot, viii.

191. For details of the Giant Power proposal, see Christie.

192. Pinchot, viii.

193. See, for example, the Museum of Modern Art exhibition *Counter Space: Design and the Modern Kitchen* (September 15, 2010–May 2, 2011).

194. The concept of the historic bloc is from Italian Marxist Antonio Gramsci, who argued that the creation of such an alliance of class forces was necessary in order for a revolutionary party to win hegemony.

195. On the early, fragmented history of the grid, see Gretchen Bakke, *The Grid: The Fraying Wires Between Americans and Our Energy Future* (New York: Bloomsbury, 2016), 40.

196. Bakke, 52.

197. Bakke, 61–70.

198. John Stuart Mill, *Principles of Political Economy, with Some of Their Applications to Social Philosophy* (New York: Little, Brown, 1848), 173.

199. Richard Ely, *Problems of To-Day: A Discussion of Protective Tariffs, Taxation, and Monopolies* (New York: Thomas Y. Crowell, 1888).

200. Adam Plaiss, "From Natural Monopoly to Public Utility: Technological Determinism and the Political Economy of Infrastructure in Progressive-Era America," *Technology and Culture* 57, no. 4 (October 2016): 806–30.

201. Richard Ely, *An Introduction to Political Economy* (New York: Chattaqua Press, 1889), 254.

202. Richard Ely, *Outlines of Economics*, 304.

203. Henry Call, *The Coming Revolution*, 173.

204. Plaiss, 819.

205. H. D. Lloyd, "The Story of a Great Monopoly," *The Atlantic*, March 1881.

206. Richard White, "For Tech Giants, a Cautionary Tale from 19th Century Railroads on the Limits of Competition," *The Conversation*, March 6, 2018.

207.   Christie, 487.

208.   Cooke to Pinchot, March 10, 1924, Cooke Papers, box 35, file 391.

209.   Pinchot, viii.

210.   Bob Johnson, *Carbon Nation*, 114.

211.   Johnson, 114.

212.   Pinchot, xi.

213.   Pinchot, xi.

214.   Pinchot, ix.

215.   Richard Rudolph and Scott Ridley, *Power Struggle: The Hundred-Year War over Electricity* (New York: Harper & Row, 1986), 38.

216.   Samuel Insull, "President's Address," *National Electric Light Association, Proceedings* (1898), 25.

217.   Insull, 24.

218.   Insull, 25.

219.   Insull, 27.

220.   Rudolph and Ridley, 39.

221.   Quoted in Rudolph and Ridley, 40.

222.   Rudolph and Ridley, 40.

223.   Pinchot, x.

224.   Pinchot, x.

225.   Pinchot, xi.

226.   Basil Manly, *Pennsylvania Grange News*, August 1923, 3–4.

227.   Editorial, *Pennsylvania Grange News*, May 1923, 8.

228.   Pinchot, xii.

229.   For a critique of technological determinism, see David E. Nye, *Consuming Power: A Social History of American Energies* (Cambridge, MA: MIT Press, 1999), 2–5.

230.   Alexander Dow, "Exploding Some Myths on Superpower and "Giant Power," *NELA Bulletin* XI (1925), 165.

231.   Carl D. Thompson, *Confessions of The Power Trust: A Summary of the Testimony Given in the Hearings of the Federal Trade Commission on Utility Corporations Pursuant to Resolution No. 83 of the Unites States Senate* (New York: E. P. Dutton, 1932).

232.   Thompson, 269.

233.   Thompson, 284–85.

234. Thompson, 316–17.

235. "Techniques Available to the Living Newspaper Dramatist," Federal Theater Project booklet (1938).

236. Federal Theatre Project, *Federal Theatre Plays*, edited by Pierre De Rohan. (New York: Da Capo, 1973).

237. *Power*, 17.

238. Richard Rudolph and Scott Ridley, *Power Struggle: The Hundred-Year War over Electricity* (New York: Harper & Row, 1986), 35.

239. *Power*, 22.

240. *Power*, 40–43.

241. *Power*, 52.

242. *Power*, 52.

243. *Power*, 53.

244. *Power*, 38.

245. See Matt Huber, "A Real Green New Deal Means Class Struggle," *Jacobin*, March 21, 2019.

246. *Power*, 63.

247. *Power*, 71.

248. *Power*, 86.

249. For a criticism of liberal hopes for justice from the Supreme Court, see Rob Hunter, "Waiting for SCOTUS," *Jacobin*, June 1, 2014.

250. Quoted in Waldemar Kaempffert, "Power for the Abundant Life," *New York Times*, August 23, 1936.

251. Ibid.

252. Ibid.

253. Kaempffert, "Power for the Abundant Life."

254. On the history of the REA, see D. Clayton Brown, *Electricity for Rural America: The Fight for the REA* (Westport, CT: Greenwood, 1980); Kiran Klaus Patel, *The New Deal: A Global History* (Princeton, NJ: Princeton University Press, 2017), 218–25; and David Nye, *Electrifying America: Social Meanings of a New Technology, 1880 – 1940* (Cambridge, MA: MIT Press, 1992), 292–304.

255. Brown, *Electricity for Rural America*, 75, 122–29.

256. Nathan Schneider, "Energy Democracy and the Billion-Dollar Co-Op," *The Nation*, May 8, 2017.

257. See Denise Fairchild and Al Weinrub, *Energy Democracy: Advancing Equity in Clean Energy Solutions* (Washington, DC: Island Press, 2017).

258. On the history of Leftist theorization of the state, see Nicos Poulantzas, *State, Power, Socialism* (New York: Verso, 2014), 254.

259. Poulantzas, 255.

260. Poulantzas, 256.

261. See, for example, John Holloway, *Change the World Without Taking Power* (New York: Pluto Press, 2002); and Marina Sitrin, ed., *Horizontalism: Voices of Popular Power in Argentina* (Oakland, CA: AK Press, 2006).

262. See Nathan Schneider's "Energy Democracy" for an account of this awakening.

263. Jackson Koeppel, "Organizing for Energy Democracy in the Face of Austerity," *Transnational Institute*, April 14, 2019.

264. Koeppel, "Organizing for Energy Democracy."

265. Ashley Dawson, "Introduction: New Enclosures" *new formations* 69 (Winter 2009).

266. See "Introduction to the New Enclosures" in Silvia Federici, *Re-enchanting the World: Feminism and the Politics of the Commons* (Oakland, CA: PM Press, 2019); and David Harvey, *The New Imperialism* (New York, Oxford University Press, 2003).

267. Harvey, *The New Imperialism*, 144.

268. Thomas More, *Utopia* (New York, Adamant Media, 2005), 19.

269. This is one of the key interventions of David Harvey's *The New Imperialism*, but his arguments flow out of a broader resistance to enclosure of the commons in the global justice movement of the 1990s and after.

270. Jim Schultz, "The Politics of Water in Bolivia," *The Nation*, January 28, 2005.

271. William Finnegan, "Leasing the Rain," *New Yorker*, March 21, 2002.

272. Oscar Olivera, *¡Cochabamba!: Water War in Bolivia* (New York: South End Press, 2004).

273. Pierre Dardot and Christian Laval, *Common: On Revolution in the 21ˢᵗ Century* (New York: Bloomsbury, 2019), 7.

274. For a discussion of popular sovereignty and the critique of liberal democracy, see Gianpaolo Baiocchi, *We, the Sovereign* (Medford, MA: Polity, 2018), 22.

275. On the social movement party, see Baiocchi, *We, the Sovereign*, 35–71.

276. See, for example, Hilary Wainwright, *Reclaim the State: Experiments in Popular Democracy* (New York: Verso, 2003).

277. Federici, *Re-enchanting the World*.

278. Koeppel, "Organizing for Energy Democracy."

279. Shane Brennan, "Visionary Infrastructure: Community Solar Streetlights in Highland Park," *Journal of Visual Studies* 16, no. 2 (2017): 167–89.

280. Koeppel, "Organizing for Energy Democracy."

281. Ibid.

282. Denise Fairchild and Al Weinrub, "Introduction," in *Energy Democracy: Advancing Equity in Clean Energy Solutions*, edited by Denise Fairchild and Al Weinrub (Washington, DC: Island Press, 2017), 13.

283. Ibid.

284. Ibid.

285. Cecilia Martinez, "From Commodification to the Commons: Charting the Pathway for Energy Democracy," in Fairchild and Weinrub, *Energy Democracy*, 27.

286. Ibid, 27.

287. Ibid, 27.

288. Ibid, 29.

289. Endnote to Solarity introduction.

290. Garrett Hardin, "The Tragedy of the Commons," *Science* 162, no. 3859 (December 13, 1968): 1243–48.

291. Ibid, 1244.

292. A. N. Whitehead, quoted in Hardin, "The Tragedy of the Commons," 1244.

293. Rob Nixon, "Neoliberalism, Genre, and the 'Tragedy of the Commons,'" *PMLA* 127, no. 33 (May 2012): 593–99.

294. Ibid, 593.

295. Hardin, "The Tragedy of the Commons," 1244.

296. Ibid.

297. Southern Poverty Law Center, "Garrett Hardin," www.splcenter.org.

298. Susie Cagle, "Bees, Not Refugees: The Environmentalist Roots of Anti-Immigrant Bigotry," *Guardian*, August 16, 2019, www.theguardian.com.

299. Nicholas Kulish and Mike McIntire, "The New Nativists: Why an Heiress Spent Her Fortune Trying To Keep Immigrants Out," *New York Times*, August 14, 2019.

300. For a summary of the flaws in Hardin's argument, see Guido Ruivenkamp and Andy Hilton, "Introduction," in Guido Ruivenkamp and Andy Hilton, eds., *Perspectives on Commoning: Autonomist Principles and Practices* (London: Zed Books, 2017), 2.

301. Hardin himself subsequently sought to roll back his argument by stating that "a managed commons [. . .] is not automatically subject to the same fate as an unmanaged commons." Thus even Hardin admitted that his initial assumption that common property was the problem was incorrect; instead, the destruction of the commons was wrought by private property (cattle) and selfish individualist behavior. See Garrett Hardin, "The Tragedy of the *Unmanaged* Commons: Population and the Disguises of Providence," in Robert V. Andelson, ed., *Commons Without Tragedy: Protecting the Environment from Overpopulation – A New Approach* (London: Shepheard-Walwyn, 1991), 162–85.

302. Elinor Ostrom, *Governing the Commons: The Evolution of Institutions for Collective Action* (New York: Cambridge University Press, 1990), 5.

303. Ibid, 8.

304. Ibid, 10.

305. Ibid, 20.

306. Ibid, 21.

307. This is the key argument and purpose of many works of radical political philosophy, from Hardt and Negri's *Common Wealth* to Dardot and Laval's *Common*.

308. Terence Daintith, "The Rule of Capture: The Least Worst Property Rule for Oil and Gas," in Aileen McHarg, Barry Barton, Adrian Bradbrook, and Lee Godden, eds., *Property and the Law in Energy and Natural Resources* (New York: Oxford University Press, 2010), 140–58.

309. This legal rule was first recognized judicially in 1886 in *Wood County Petroleum Co v. West Virginia Transportation Co.* For more on this history, see Daintith, 140.

310. Daintith, "The Rule of Capture," 147.

311. Matthew Huber, *Lifeblood: Lifeblood: Oil, Freedom, and the Forces of Capital* (Minneapolis: University of Minnesota Press, 2013), 44.

312. Ibid, 45.

313. In his overview of the rule of capture, Terence Daintith notes that the trend toward overproduction produced by the law of capture has been the norm in the United States. See Daintith, "The Rule of Capture," 141.

314. For a detailed discussion of the overexploitation of the Black Giant, see Huber, *Lifeblood*, 48–50.

315. Cited in Huber, 48.

316. Cited in Huber, 48.

317. Cited in Huber, 51.

318. Huber, 54.

319. Elinor Ostrom, *Governing the Commons: The Evolution of Institutions of Collective Action* (New York: Cambridge University Press, 1990).

320. "Who We Are," *The Solar Commons*, www.solarcommons.org.

321. "What are Solar Commons?," *The Solar Commons*, www.solarcommons .org.

322. "Solar Commons: How Does It Work?," *The Solar Commons*, www .solarcommons.org.

323. Kathryn Milun and Matthew Grimley, *Solar Commons: Designing Community Trust Solar Ownership for Social Equity* (Spring 2017), available at www.solarcommons.org.

324. Milun and Grimley, *Solar Commons*, 12.

325. Milun and Grimley, *Solar Commons*, 13.

326. Ibid.

327. On the commons and the *Magna Carta*, see Peter Linebaugh, *The Magna Carta Manifesto: Liberties and Commons for All* (Berkeley: University of California Press, 2009). On indigenous struggles and on social reproduction, see Silvia Federici, *Re-enchanting the World: Feminism and the Politics of the Commons* (Brooklyn: Autonomedia, 2019), 78–84, and 175-186.

328. Ruivenkamp and Hilton, "Introduction," 4.

329. Milun and Grimley, *Solar Commons*, 15.

330. Ibid.

331. For a critique of the energy-as-commodity paradigm, see Cecilia Martinez, "From Commodification to the Commons: Charting the Pathway for Energy Democracy," in Denise Fairchild and Al Weinrub, eds., *Energy Democracy: Advancing Equity in Clean Energy* (Washington, DC: Island Press, 2017), 22–36.

332. Denise Fairchild and Al Weinrub, "Introduction," in Denise Fairchild and Al Weinrub, eds., *Energy Democracy: Advancing Equity in Clean Energy* (Washington, DC: Island Press, 2017), 13.

333. Mitchell, 5.

334. Mitchell, 38.

335. Caffentzis, 63.

336. This explanation of capital, machines, labor, and nature is indebted to Andreas Malm's autonomy-inspired account in *The Progress of this Storm*, 197–214.

337. Jan Ellen Spiegel, "Another $1.2 billion substation? No thanks, says utility, we'll find a better way," *Inside Climate News*, April 4, 2016.

338. Lawrence Orsini, interview, January 17, 2018.

339. Jesse Morris, interview, January 16, 2018.

340. Jesse Morris, interview. See also Merlinda Andoni et al., "Blockchain Technology in the Energy Sector: A Systematic Review of Challenges and Opportunities," *Renewable and Sustainable Energy Reviews* 100 (February 2019): 143–74.

341. Garry Golden, interview, January 12, 2018. See also Diane Cardwell, "Solar Experiment Lets Neighbors Trade Energy Among Themselves," *New York Times*, March 13, 2017.

342. Garry Golden, interview.

343. Fairfield and Weinrub, "Introduction," in Fairchild and Weinrub, Eds., *Energy Democracy*, 6.

344. Michael Hardt and Antonio Negri, *Commonwealth* (Harvard University Press, 2009), ix-x.

345. David Harvey, *Rebel Cities: From the Right to the City to the Urban Revolution* (New York: Verso, 2012), 73.

346. Maria Gallucci, "Energy Equity: Bringing Solar Power to Low-Income Communities," *Yale Environment 360*, April 4, 2019.

347. Gallucci, "Energy Equity."

348. James Angel, "Towards an Energy Politics In-Against-and-Beyond the State: Berlin's Struggle for Energy Democracy," *Antipode* 49, no. 33 (June 2017): 557–76. See also Gianpaolo Baoicchi, *We, the Sovereign* (Medford, MA: Polity Press, 2018).

349. Nicos Poulantzas, *State, Power, Socialism* (New York: Verso, 1978), 258.

350. Ibid.

351. Denise Fairchild and Al Weinrub, "Introduction," 10.

352. Ellie O'Byrne, "An Exploration of Oil as the Devil's Excrement," *Irish Times* (April 4, 2017).

353. Antti Salminen and Tere Vadén, *Energy and Experience: An Essay in Nafthology* (MCM¹, 2015), 2.

354. Ibid, 21.

355. Rhys Williams, "'This Shining Confluence of Magic and Technology:' Solarpunk, Energy Imaginaries, and the Infrastructures of Solarity," *Open Library of Humanities* 5, no. 1 (2019): 14.

356. "Summer Project," C. Arseneault and B. Pierson, eds., *Wings of Renewal: A Solarpunk Dragon Anthology* (Florida: Incandescent Phoenix Books, 2014).

357. Pierre Dardot and Christian Laval, *Common: On Revolution in the 21st Century* (New York: Bloomsbury, 2019), 5.

358. Ibid, 6.

359. See "Solarity: After Oil School 2," a workshop organized by the Petrocultures Research Group, www.afteroil.ca/solarity.

360. Martinez, "From Commodification to the Commons," 27.

361. Ibid, 28.

362. Ibid.

363. Ibid.

364. Ibid, 29.

365. Ibid, 28.

366. Kathryn Milun, *The Political Uncommons: The Cross-Cultural Logic of the Global Commons* (Burlington, VT: Ashgate, 2011).

367. Kathryn Milun, interview, August 16, 2019.

368. Ibid.

369. Milun, *The Political Uncommons*, 1. On narrative and national imaginaries, see Benedict Anderson, *Imagined Communities: Reflections on the Origins and Spread of Nationalism* (New York: Verso, 1998).

370. Dardot and Laval, *Common*, 17.

371. Ibid.

372. Milun, *The Political Uncommons*, 8.

373. Ibid, 11.

374. Denise Fairchild and Al Weinrub, "Introduction," 12.

375. See Vaclav Smil, *Power Density: A Key to Understanding Energy Sources and Uses* (Cambridge, MA: MIT Press, 2015).

376. John Van Zalk and Paul Behrens, "The Spatial Extent of Renewable and Non-Renewable Power Generation," *Energy Policy* (December 2018): 83–91.

377. Leiden University, "Renewable Energy Sources Can Take Up to 1000 Times More Space Than Fossil Fuels," *Phys.org*.

378. For a discussion of the political and ecological implications of power density, see Troy Vettese, "To Freeze the Thames: Natural Geo-Engineering and Biodiversity," *New Left Review* 111 (May–June 2018): 63-86.

379. Ibid.

380. Alexander Dunlap, "Wind Energy: Toward a 'Sustainable Violence' in Oaxaca," *NACAL Report on the Americas* 49, no. 4 (2017): 483–88. See also Cymene Howe, "Aeolian Extractivism and Community Wind in Southern Mexico," *Public Culture* 28, no. 279 (2016): 215–35; Sofia Avila-Calero, "Contesting Energy Transitions: Wind Power and Conflicts in the Isthmus of Tehuantepec," *Journal of Political Ecology* 24, no. 1 (September 27, 2017): 992–1012; and Ezequiel Zárate-Toledo, Rodrigo Patiño, and Julia Fraga, "Justice, Social Exclusion and Indigenous Opposition: A Case Study of Wind Energy Development on the Isthmus of Tehuantepec, Mexico," *Energy Research & Social Science* 54 (August 1, 2019): 1–11.

381. Dunlap, 485.

382. Victoria Burnett, "La Ventosa Journal: Mexico's Wind Farms Brought Prosperity, But Not for Everyone," *New York Times*, July 26, 2016.

383. Dunlap 486.

384. Subcomandante Marcos, quoted in Dunlap 483.

385. Environmental Justice Atlas, "Corporate Wind Farms in Ixtepec vs Community's Initiative, Oaxaca, Mexico," March 29, 2017, www.ejatlas .org.

386. Sergio Oceransky, "Fighting the Enclosure of Wind: Indigenous Resistance to the Privitization of Wind Resources in Southern Mexico," in Kolya Abramsky, ed., *Sparking A Worldwide Energy Revolution Social Struggles In The Transition To A Post-Petrol World* (Oakland, CA: AK Press, 2010), 505–22.

387. Martinez, "From Commodification to the Commons," 29.

388. Noah Goldberg, "Brooklynites Who Lost Power Ask Con Edison: 'Why Us?'" *Brooklyn Eagle*, July 22, 2019.

389. Will Wade, Brian Eckhouse, and Henry Goldman, "NYC Mayor Suggests ConEd Takeover After Heat Forces Shutdown," *Bloomberg*, July 21, 2019.

390. "Con Ed and National Grid Talking Points and Research," Ecosocialist Working Group, Democratic Socialists of America, July 22, 2019.

391. "Borrowing at High Interest to Pay Unaffordable Utility Bills," *New York's Utility Project*, October 10, 2012, www.utilityproject.org.

392. "Con Ed and National Grid Talking Points and Research," Ecosocialist Working Group, Democratic Socialists of America, July 22, 2019.

393. Corey Kilgannon, "Manhole Fires and Burst Pipes: How Winter Wreaks Havoc on What's Beneath NYC," *New York Times*, February 21, 2019.

394. Dan Rivoli, "Con Ed asks New Yorkers to Cough Up $695M in Rate Hikes," *Daily News*, January 31, 2019.

395. Shoshona Zuboff, *The Age of Surveillance Capitalism: The Fight for a Human Future at the New Frontier of Power* (New York: PublicAffairs, 2019).

396. New York City Department of Health, "Community Health Profiles 2015: Queens Community District 1," www1.nyc.gov.

397. "America's Dirtiest Power Plants," Environment New York Research and Policy Center, September 2014, www.environmentnewyorkcenter.org.

398. Fabio Caiazzo et al., "Air Pollution and Early Deaths in the United States," *Atmospheric Environment* 79 (November 2013): 198–208.

399. "We Must Stop New York's 'Peaker Plants' Choking Marginalized Communities," *New York Lawyers for the Public Interest* (February 11, 2019), www.nylpi.org.

400. Clayton Guse, "Giant Electric Battery Set Will Help Curb Ravenswood Plant Pollution in Queens, State Says," *Daily News*, October 17, 2019.

401. "Environmental Impacts of Natural Gas," Union of Concerned Scientists, June 19, 2014.

402. On the Sunset Park Solar Co-op, see Lourdes Pérez-Medina and Elizabthe Yeampierre, "The People's Power," *Urban Omnibus*, April 10, 2019.

403. "Renewable Energy," Con Edison, www.coned.com.

404. "Public Utilities Under Public Control," Public Power NYC, www.publicpower.nyc.

405. "Forming a Public Power Utility," Public Power, www.publicpower.org.

406. Catherine Morehouse, "Chicago Considers Municipalizing ComEd," *Utilitydive* (July 25, 2019); and Kevin Stark, "Power Switch: SF Builds Case for Pushing Out PG&E" *SFPublic Press*, June 18, 2019.

407. Amber Ruther and Steve Fox, interview, August 15, 2019.

408. Ibid.

409. Energy Information Administration, *Short-Term Energy Outlook*, November 13, 2019.

410. Andrew Cumbers, "Public Ownership as Economic Democracy" in Andrew Cumbers et al., "Public Ownership and Alternative Political Horizons," *Soundings* 64 (Winter 2017): 83–104. For further important arguments in the need to reclaim and reconstitute public ownership, see Andrew Cumbers, *Reclaiming Public Ownership: Making Space for Economic Democracy* (New York: Zed Books, 2012).

411. "Renewables Generated a Record 65 Percent of Germany's Electricity Last Week," *Yale Environment 360 Digest*, March 13, 2019.

412. Craig Morris and Anre Jungjohann, *Energy Democracy: Germany's Energiewende to Renewables* (New York: Palgrave Macmillan, 2016), ix.

413. Morris and Junghohann, x.

414. Ibid, 4.

415. Gretchen Bakke, *The Grid: The Fraying Wires Between Americans and Our Energy Future* (New York: Bloomsbury, 2016), 87, 93–104.

416. Ibid, 103.

417. IRENA, *30 Years of Policies for Wind Energy: Lessons from Denmark* (January 2013), 59.

418. For a detailed discussion of this global upsurge of protest, see George Katsiaficas, *The Subversion of Politics: European Autonomous Social Movements and the Decolonization of Everyday Life* (Oakland, CA: AK Press, 2006).

419. Ibid, 5.

420.   Ibid, 3–4.

421.   Ibid, 81.

422.   Ibid, 86.

423.   Ibid, 7.

424.   Ibid, 11.

425.   Morris and Junghohann, 59.

426.   Ibid.

427.   Ibid, 194.

428.   Sean Sweeney and John Treat, *Preparing a Public Pathway: Confronting the Investment Crisis in Renewable Energy*, TUED Working Papers #10, November 2017. [QU: Edits acceptable?]

429.   Ibid, 18.

430.   Morris and Junghohann, 8.

431.   Henner Busch, interview (February 13, 2017).

432.   Renuka Rayasam, "A Power Grid of Their Own: German Village Becomes Model for Renewable Energy," *Der Spiegel*, March 9, 2012.

433.   Sweeney and Treat, *Preparing a Public Pathway*, 19.

434.   Ibid, 20.

435.   Beth Gardiner, "For Europe's Far-Right Parties, Climate is the New Battleground," *Yale Environment 360*, October 29 2019.

436.   Ibid.

437.   Sweeney and Treat, *Preparing a Public Pathway*, 25.

438.   Ibid.

439.   Christopher Rasch, "Energiewende Retten," *Greenpeace Energy*, November 29, 2019.

440.   For a detailed account of Berlin's water wars, see Andrea Muehlebach, "Towards a Social Infrastructure," *e-flux*, n.d.

441.   Transnational Institute, *Reclaiming Public Services: How Cities and Citizens Are Turning Back Privatization*, June 23, 2017.

442.   Nora Rocholl and Ronan Bolton, "Berlin's Electricity Distribution Grid: An Urban Energy Transition in a National Regulatory Context," *Technology Analysis and Strategic Management* 28, no. 10 (2016): 1188.

443.   For a history of the Energietisch, see www.berliner-energietisch.net/ english-information.

444. Conrad Künzer and Sören Becker, *Energy Democracy in Europe: A Survey and Outlook*, Rosa Luxemburg Foundation, Belgium. [QU: Edits acceptable?]

445. Michael Efler, interview, February 3, 2017.

446. Luise Neumann-Cosel, interview, February 2, 2017.

447. Ibid.

448. James Angel, "Toward An Energy Politics In-Against-And-Beyond the State: Berlin's Struggle for Energy Democracy," *Antipode* 49, no. 3 (June 2017): 557–76.

449. Luise Neumann-Cosel, interview.

450. Michael Efler, interview, February 3, 2017.

451. Angel, "Toward An Energy Politics In-Against-And-Beyond the State."

452. Nicos Poulantzas, *State, Power, Socialism* (New York: Verso, 2014), 252.

453. Ibid, 254.

454. Ibid, 255.

455. John Holloway, *Change the World Without Taking Power: The Meaning of Revolution Today* (New York: Pluto Press, 2002).

456. See Marina Sitrin, *Horizontalism: Voices of Popular Power in Argentina* (Oakland, CA: AK Press, 2006), 4.

457. Poulantzas, 256.

458. Angel, 563.

459. Oliver Powalla, interview, February 23, 2018.

460. Ibid.

461. Berlin Energietisch, Press Release of 5 March 2019, www.berliner-energietisch.net.

462. Oliver Powalla, interview, February 23, 2018.

463. "The City of Hamburg is Rebuying and Transforming Its District Heating System," *LowTemp* (February 8, 2018).

464. On the Green New Deal, see Kate Aronoff, Alyssa Battistoni, Daniel Aldana Cohen, and Thea Riofrancos, *A Planet to Win: Why We Need a Green New Deal* (New York: Verso, 2019).

465. Sweeney and Treat, *Preparing a Public Pathway*, 50.

466. T. X. Watson, "The Boston Hearth Project" in Phoebe Wagner and Brontë Christopher Wieland, *Sunvault: Stories of Solarpunk and Eco-Speculation* (Nashville, NT: Upper Rubber Boot Books, 2017), 16.

467. Watson, "The Boston Hearth Project," 24.

468. Ibid.

469. Rhys Williams, "'This Shining Confluence of Magic and Technology:' Solarpunk, Energy Imaginaries, and the Infrastructures of Solarity," *Open Library of Humanities* 5, no. 1 (2019): 23.

470. Ibid, 24.

471. Ibid, 14.

472. Ibid, 14.

473. Ibid, 20.

474. Cited in Jason Hickel, "The Limits of Clean Energy," *Foreign Policy* (September 6, 2019).

475. World Bank, *The Growing Role of Minerals and Metals for a Low Carbon Future* (June 2017), xii.

476. Hickel, "The Limits of Clean Energy."

477. Alanah Torralba, Tadzio Müller, and Elis Soldatelli, "Pumping the Brakes on E-cars: Unmasking the Fantasy of Green Capitalism," *Rosa Luxemburg Stiftung* (December 11, 2019).

478. Stan Cox and Paul Cox, "100% Wishful Thinking: The Green Energy Cornucopia," *Counterpunch*, September 14, 2017.

479. See, for example, "What Does It Mean to Achieve a Fast and Just Transition to 100% Renewable Energy?" www.gofossilfree.org/renewable-energy.

480. United Nations, *Sustainable Development Goals #7*, "Affordable and Clean Energy," www.un.org.

481. Andrew Sayer, *Why We Can't Afford the Rich* (Chicago: University of Chicago Press, 2014).

482. Stefanos Chen, "At $238 Million, It's the Highest-Price Home in the Country," *New York Times*, January 23, 2019.

483. David Graeber, *Bullshit Jobs: A Theory* (Simon & Schuster, 2018).

484. See Lizabeth Cohen, *A Consumer's Republic: The Politics of Mass Consumption in Postwar America* (New York: Knopf, 2003); and Dolores Hayden, *Building Suburbia: Green Fields and Urban Growth, 1820-2000* (New York: Vintage, 2004).

485. Matthew Nitch Smith, "The Number of Cars Will Double Worldwide by 2040," *Business Insider*, April 20, 2016.

486. Mike Berners-Lee and Duncan Clark, "What's the Carbon Footprint of . . . a New Car," *Guardian*, September 23, 2010.

487. Ibid.

488. See Ken Hiltner, *Forward to Nature: Writing a New Environmental Era* (New York: Routledge, 2019).

489. Ibid.

490. See Matthew Huber, "Five Principles of a Socialist Climate Politics," *The Trouble* (August 16, 2018).

491. Stan Cox and Paul Cox, "100% Wishful Thinking."

492. Richard Smith, "Capitalism and the Destruction of Life in Earth: Six Theses on Saving the Humans," *Real-World Economics Review* (2013): 136.

493. International Energy Association, *Population Without Access to Electricity Falls Below 1 Billion*, October 30, 2018, www.iea.org.

494. Rebecca Rewald, "Does Providing Energy Access Improve the Lives of Women and Girls? Sort of," *Oxfam*, June 6, 2017, www.politicsofpoverty.oxfamamerica.org.

495. Isabel Hilton, "How China's Belt and Road Initiative Threatens Global Climate Progress," *Yale Environment 360*, January 3, 2019.

496. For an attack on all criticisms of the capitalist doctrine of inexorable growth, see Leigh Phillips, *Austerity Ecology and the Collapse-Porn Addictions: A Defense of Growth, Progress, Industry and Stuff* (London: Zero Books, 2015).

Born in South Africa during the apartheid era, Ashley Dawson is currently Professor of Postcolonial Studies at the Graduate Center, City University of New York and the College of Staten Island. His previous books include *Extreme Cities: The Peril and Promise of Urban Life in the Age of Climate Change* and *Extinction: A Radical History*. A member of the Social Text Collective and the founder of the CUNY Climate Action Lab, he is a long-time climate justice activist.